Events in Over **29 countries**! Jo...

JANUARY:
- Bangkok
- ★ Bonita Springs

FEBRUARY:
- Cairo
- ★ Kuwait
- Caguas
- ★ Elkhorn
- Roanoke
- ★ Jacksonville
- Hattiesburg

MARCH:
- ★ Rhur
- Sachsen
- ★ Malta
- The Big Muddy
- ★ Lafayette, Louisiana
- Cache Valley
- ★ Lynchburg
- Viña

APRIL:
- Pilsen
- ★ Zadar
- Lille
- ★ Berlin
- Hsinchu
- ★ Connecticut
- Miami
- ★ Meridian
- Pioneer Valley
- ★ Wichita
- Martinsville
- ★ Cape Cod
- Tyler
- ★ Loma Linda
- Mercer-Bucks
- ★ Asheville
- Bowling Green
- ★ Cranberry Township
- South Bend
- ★ Bloomsburg
- Greater Lafayette, Indiana
- ★ Lake County

MAY:
- Vienna
- ★ Saskatoon
- Austin
- ★ Dugo Selo
- Kyoto
- ★ Luxembourg
- Alentejo
- ★ Burlington
- Gilroy
- ★ North Little Rock
- Columbia, South Carolina
- ★ Hawthorne
- Stark
- ★ Wayne County
- Yucatán
- ★ Southwest Michigan

JUNE:
- Prague
- ★ Trieste
- Manila
- ★ Long Island
- South Shore
- ★ Sheboygan
- Galicia
- ★ NoVA
- Delft

JULY:
- ★ Edmonton
- Windhoek
- ★ Detroit
- Coeur D'Alene
- ★ Kingsport
- Tampa
- ★ Glasgow

AUGUST:
- Tokyo
- ★ Seoul
- Wellington
- ★ Alameda
- Tulsa
- ★ Asunción

SEPTEMBER:
- Hannover
- ★ Lisbon
- Shanghai
- ★ Moscow
- Eindhoven
- ★ Milwaukee
- Louisville
- ★ Des Moines
- Shreveport-Bossier
- ★ Rogue Valley

OCTOBER:
- Zagreb
- ★ Rome
- Warsaw
- ★ The Ozarks
- Downtown Columbia
- ★ Fredonia
- Philadelphia
- ★ Atlanta
- Twin Tiers
- ★ Taipei
- Baton Rouge
- ★ Colorado Springs
- Syracuse
- ★ Barcelona

NOVEMBER:
- Paris
- ★ Sindelfingen
- Jalisco
- ★ Cleveland
- Orlando
- ★ Rochester
- Pensacola
- ★ York County
- Madison
- ★ Mexicali
- Bilbao

makerfaire.com

CONTENTS

ON THE COVER:
Keep your voice assistant and smart home data to yourself with private-by-design tools. Photograph by Mark Madeo.

Top right: This retro arcade briefcase provides go-anywhere fun. Photo: Tyler Capps; Street Fighter II image: Capcom

36

PROJECTS

SKILL BUILDER

TOOLBOX

OVER THE TOP

Mark Wade), Bjørn Karmann and Tore Knudsen, Brett McAfee, Argus McIntyre, Justin Lemire-Elmore

46

22

08

64

Make:

"The best way to secure data is never to collect it in the first place." —Cory Doctorow

PRESIDENT
Dale Dougherty
dale@make.co

VP, PARTNERSHIPS
Todd Sotkiewicz
todd@make.co

EDITORIAL

EXECUTIVE EDITOR
Mike Senese
mike@make.co

SENIOR EDITORS
Keith Hammond
keith@make.co
Caleb Kraft
caleb@make.co

PRODUCTION MANAGER
Craig Couden

CONTRIBUTING EDITOR
William Gurstelle

CONTRIBUTING WRITERS
Stan Adermann, Vern Adermann, Brian Bunnell, Tyler Capps, Mara Capron, Amanda Formaro, Kris Gesling, Kathy Giori, Saul Griffith, Bjørn Karmann, Christine Knobel, Justin Lemire-Elmore, Forrest M. Mims III, Samer Najia, David Perry, Kevin Purdy, Kathy Reid, Mark Schiess, Philip Schmidt, Lex Sugden

DESIGN & PHOTOGRAPHY

CREATIVE DIRECTOR
Juliann Brown

CONTRIBUTING ARTISTS
Mark Madeo

MAKE.CO

ENGINEERING MANAGER
Alicia Williams

WEB APPLICATION DEVELOPER
Rio Roth-Barreiro

GLOBAL MAKER FAIRE

MANAGING DIRECTOR, GLOBAL MAKER FAIRE
Katie D. Kunde

MAKER RELATIONS
Sianna Alcorn

GLOBAL LICENSING
Jennifer Blakeslee

MARKETING

DIRECTOR OF MARKETING
Gillian Mutti

OPERATIONS

OPERATIONS DIRECTOR
Cathy Shanahan

ACCOUNTING MANAGER
Kelly Marshall

OPERATIONS MANAGER & MAKER SHED
Rob Bullington

PUBLISHED BY

MAKE COMMUNITY, LLC
Dale Dougherty

Copyright © 2020
Make Community, LLC. All rights reserved. Reproduction without permission is prohibited. Printed in the USA by Schumann Printers, Inc.

Comments may be sent to:
editor@makezine.com

Visit us online:
make.co

Follow us:
🐦 @make @makerfaire @makershed
f makemagazine
makemagazine
▶ makemagazine
twitch.tv/make
makemagazine

Manage your account online, including change of address: makezine.com/account
866-289-8847 toll-free in U.S. and Canada
818-487-2037,
5 a.m.–5 p.m., PST
cs@readerservices.makezine.com

Make: Community

Support for the publication of Make: magazine is made possible in part by the members of Make: Community. Join us at make.co.

CONTRIBUTORS

What's your go-to for when you just want to get offline for a bit?

Kathy Giori
Stanford, CA (Your Own Private Smart Home)
I like to mountain bike up Arastradero preserve to Foothills Park (no biking) where I can hike atop a scenic ridge overlooking the SF Bay Area.

Justin Lemire-Elmore
Vancouver, Canada (Boost That Bike!)
My absolute favorite time-out is hopping in our rowboat and cruising the nooks and crannies of Vancouver harbour with my wife and baby girl.

Kathy Reid
Geelong, Australia (Open Source Voice for Makers)
I'm an avid knitter, and I love to create things with yarn.

Issue No. 72, Spring 2020. *Make:* (ISSN 1556-2336) is published quarterly by Make Community, LLC, in the months of February, May, Aug, and Nov. Make Community is located at 150 Todd Road, Suite 200, Santa Rosa, CA 95407. SUBSCRIPTIONS: Send all subscription requests to *Make:*, P.O. Box 17046, North Hollywood, CA 91615-9588 or subscribe online at makezine.com/offer or via phone at (866) 289-8847 (U.S. and Canada); all other countries call (818) 487-2037. Subscriptions are available for $34.99 for 1 year (4 issues) in the United States; in Canada: $43.99 USD; all other countries: $49.99 USD. Periodicals Postage Paid at San Francisco, CA, and at additional mailing offices. POSTMASTER: Send address changes to *Make:*, P.O. Box 17046, North Hollywood, CA 91615-9588. Canada Post Publications Mail Agreement Number 41129568. CANADA POSTMASTER: Send address changes to: Make Community, PO Box 456, Niagara Falls, ON L2E 6V2

LET'S MAKE *Online*

Take your projects to the next level with **Make: Projects**,
a new collaborative platform that significantly improves
the way you document, share and advance projects online.
Coming soon in spring 2020!

Sign Up for Updates:
www.makeprojects.com

Make: Projects

POWERED BY ● ProjectBoard

Our Pleasure!

MOVIN' ON UP

You guys have given me such a huge boost in my self-confidence as a maker and person in general, going from a kid who couldn't build a model helicopter, to a kid who built a 3D printer and computer over the past two years. I'm so grateful for all you guys do. In fact your magazine introduced me to LoRa radio, which I used with my local 4-H team to make a safety device for local farmers. Your magazine helped our team win that competition. I can't express how much *Make:* means to me. I truly hope you guys keep doing what you do and inspire more people like me.   —*Eoghan Murphy, age 14*

Kent K. Barnes
@KeNTKB

So Cool to see our buddy @JimmyDiResta being featured in the first return issue of Make Magazine! FIX OUR PLANET youtu.be/DauFg1sbpK8 via @YouTube thank you @calebkraft .

▶ **YouTube** @YouTube

♡ 1   11:04 AM - Dec 13, 2019

Mike Swimm
@mikeswimm

I am trying to focus on the positive tonight.

I'm thankful that @make magazine has found a way to keep rolling, and thrilled that they have moved back to the original format.

Also, Make is an excellent gift for any curious tinkerers in your life. readerservices.makezine.com/mk/default.aspx

♡ 23   5:28 PM - Dec 12, 2019

---

sweet maria's coffee ✓
@sweetmarias

Have yall seen Sweet Maria's customer Larry Cotton's latest genius creation in Make Magazine? Larry provided step by step instructions for anyone who dares to give this build a try. bit.ly/2sdReX6 📷Make:Magazine/Larry Cotton

♡ 7   3:49 PM - Dec 10, 2019

  See sweet maria's coffee's other Tweets

Can't Hide These Spyin' Eyes

by Mike Senese, Executive Editor

This past November, I traveled to Guadalajara, Mexico, for the inaugural Maker Faire Jalisco. The show was fantastic, combining cutting-edge technologies with regional traditions, all set in a gorgeous park in the center of the city. Robotic exoskeletons and person-carrying quadcopters alongside *Día de los Muertos* symbolism — a wonderful melding of worlds and a great showcase of what makes that area so vibrant.

At the end of the Friday kickoff, as I prepared to head to my hotel, I started to wonder what was happening back home. Was my wife picking up my son from kindergarten? Were they out for lunch? Visiting her mom? I instinctually raised my wrist and tapped my watch. The screen changed to a map, with an orange dot showing her exact location 1,600 miles away. It gave me a warm feeling, followed by a slight amount of creepiness.

When we had our son, my wife and I agreed to keep the location sharing function activated on our devices. It's been useful for knowing when the other should arrive for dinner, to help rendezvous at events, and, when one of us travels, to provide a comforting moment of connection to home. But the ease and accuracy at which it operates certainly stretches the boundaries of privacy. It provides no indication that one of us is checking on the other. You can set alerts for when the other leaves their location, or gets close to yours. The convenience provided is countered by an inability to surprise one another, for better or worse. It reduces the autonomy that's part of being an adult.

With this issue, we jump into the conversation around tech privacy and security, a significant and growing discussion over the past couple years. Our devices have more capable eyes, ears, and brains than ever before, and with them come ample opportunities for individuals and companies to get pretty darn shady, both with and without our consent. In the following pages we will help show you the options for controlling your environment with the latest tech while still keeping your privacy shielded. And as always, I'm eager to hear about your own projects in this area — email me anytime about your undertakings: mike@make.co. ✐

MADE ON EARTH

Backyard builds from around the globe

Know a project that would be perfect for Made on Earth?
Let us know: *editor@makezine.com*

RIDING HIGH INSTAGRAM.COM/MARIADCAMINO

Maria Del Camino is, as part of its name implies, a 1959 El Camino, albeit with heavy modifications in both the "extensive" and "massive" sense. First, where you'd normally find wheels is now an industrial tracked excavator chassis. Next, the vehicle's body raises and lowers on a pair of hydraulically actuated arms, putting its passengers high above the ground. And, relating to the other part of the creation's name, you'll find the hood drilled out with tens of thousands of holes in varying diameters, creating the visage of Maria, the robot from the 1927 sci-fi film *Metropolis*.

Creator **Bruce Tombs** set on the idea to build an homage to "the unfulfilled promise of flying cars" upon his first visit to Burning Man in 2009. "The mutant vehicle culture there kicked my motorhead tendencies into high gear," he says. "I immediately began working out ideas that eventually became *Maria del Camino*." He surmises that his friends and supporters thought he was insane.

The interdisciplinary artist and architect debuted his creation a year later, and has strived to make substantial modifications every year since. The latest of those was to replace the 5-ton diesel powerplant with an electric motor that can be controlled by a smartphone. Always a crowd favorite at outdoor exhibitions, Tombs comments that this will now let *Maria Del Camino* perform indoors as well.

Indoors or out, those wanting to experience her in person will have to make their way out West. "Given her substantial weight, 5 tons, the logistics of moving her are significant," says Tombs, who recently relocated from San Francisco to Reno. "Her top speed is 1.7 miles an hour. Her travels so far have been limited to California and Nevada. She is not about going fast!" — *Mike Senese*

Angus McIntyre

SERIOUSLY, THAT'S NOT A MOON

MAKEZINE.COM/GO/DEATHSTARDAD

There are *Star Wars* fanatics — most of us — and then there's the **Powell family**.

Since 2015, the Powells have commemorated the release of each new *Star Wars* film by building a massive, iconic spacecraft from the series. In 2015, they constructed a 23-foot-diameter Death Star, made with ½" PVC conduit and specialized couplers, then skinned with an army-surplus parachute and trimmed with glowing LED strip lights. 2017's edition found them designing a more complicated wood-and-fabric structure in the shape of a 28½-foot glowing Millennium Falcon. And in late 2019 they completed the sequel trilogy with a 20-foot X-wing Fighter and 25-foot TIE Advanced, set up in a dramatic dogfight.

The builds are impressive, but their placement makes people say wow: The family (Colby, Julia, Drew, Isabel, Cameron and Ian) mounts them on the roof of their Lafayette, California home for the entire neighborhood to enjoy. It's always a serious rigging job, requiring a 70-foot crane and engineer-caliber calculations. With their Death Star, dad Colby described, "By the time you raise a 23-foot sphere 10 feet in the air, and then allow enough extra cable so that it can be disconnected from up on top of the roof without climbing a ladder, you've got to get this giant truck pretty darn close to the house to make it all work!"

Colby says he'll let the kids continue to decide on what build they may undertake moving forward. "Whatever it is, I know it will be fun to create and share." —*Mike Senese*

BIRTHDAY BOAT

KYLESCHEELE.COM/THE-VIKING-FUNERAL

As **Kyle Scheele**, a Springfield, Missouri-based professional speaker, faced his impending transition into his 30s, he stumbled upon a brilliant idea during a conversation with his mother, who asked if he was going to have a party for the milestone birthday.

"I said, 'No, mom. I'm not going to have a party,'" Scheele says. "'I'm not going to celebrate turning 30. I'm going to have a funeral to mourn the death of my 20s.'"

"She replied, 'That's weird, Kyle. Is this your idea of a birthday party theme? Are you going to have people put their presents in a coffin or something?'"

Thinking quickly, Scheele told her, "No, mom. That would be creepy. I'm not having a regular funeral for my 20s. I'm having a Viking funeral."

As soon as he said it, he knew it had to happen.

Scheele, who had been making art with cardboard for a while, suddenly faced a very tough timeline. His brilliant idea didn't arise until just a few weeks before the big day, but he powered through and finished his 8-foot-tall, 16-foot-long cardboard Viking ship with only moments to spare, before he and his friends set it aflame with Roman candles.

His friends made a documentary about the whole process and published it online. This story has gone on to inspire many people around the world, and ultimately it spawned a new project for Scheele: a bigger Viking ship full of other people's regrets (over 20,000), that he burned in mid-December. —*Caleb Kraft*

NOMCON

JUNE 5 - 7 **Eugene, Oregon**

A Conference for Leaders of Maker Organizations

Building & Supporting an Ecosystem That Sustains Us All

2020 Theme - Sustainability:

+50 Sessions, Activities, & Meetups!

Join 100s of Fellow Maker Leaders!

Hosted By:

NATION OF MAKERS

nationofmakers.us

4-STROKE STOKE INSTAGRAM.COM/JUNGEMAKERS

During a Rockler store demo by Paul Jackman, a popular woodworker known for his sense of humor and tendency to "Jackman-size" everyday items by making massive replicas of them, the **Junge family** received an unfinished penny board with a directive to complete it. Parents Deanna and Stewart and kids Brody, 13, and Colby, 10, all dove in on planning how to finish it. Scope creep quickly set in, and many hours and ideas later a family-carrying monster of a skateboard (as a tribute to Jackman) became the target.

The family owns a cabinet shop in Andover, Massachusetts, so tools and scrap materials were already at hand, but they had to do some creative engineering and tons of labor to get to the final product: a 7-foot-plus, chain-driven, off-road longboard.

"There was nothing standard about this skateboard," the family says. "We looked online but there were no off-the-shelf solutions for trucks that were going to make this the all-terrain machine it needed to be. There was a lot of trial and error."

The board is now nearly complete, but as you might imagine with a gas-powered 6.5HP engine, they're taking their time getting comfortable riding it. But good news for those who want one: they intend to share the entire build on their own YouTube channel. —*Caleb Kraft*

Courtesy Junge Family

Where Curiosity Takes You

expl⊙ratorium®

Decarbonization Begins at Home

Written by Dr. Saul Griffith

ERASE HALF YOUR CARBON FOOTPRINT — JUST BY ELECTRIFYING YOUR ROOF, YOUR HEAT, AND YOUR WHEELS

How far can a homeowner, a maker, a concerned citizen go to decarbonizing their life? Quite far!

Most climate policy sounds abstract — large-scale decisions made by people you don't know. Forget that. Here's how to start decarbonizing your own world, from your kitchen table outward, without invoking fanciful technology that doesn't exist yet. You can see for yourself right now what solving climate change successfully looks like.

Of course we can't get to a zero carbon world purely by making personal consumer decisions — we critically need government and we need an enlightened industry. But the easiest emissions to eliminate are those you directly control as a consumer: the gasoline you pour into your tank, natural gas you burn in your furnace and stovetop, the fossil fuels burned by utilities to feed your electricity.

You can broadly outline the big carbon-spewing items in your life as your transportation, your heating, your electricity, your food, your stuff, and your services (including government). Here, then, are the energy and carbon consumptions and productions you can personally affect from your kitchen table (in rough order of climate impact):

» Electrify your vehicles and transportation
» Electrify your heating and hot water
» Produce your electricity locally, or buy from renewable sources
» Eat less meat and less refrigerated food
» Buy less stuff, make it better, maintain it and make it last longer
» Decarbonize services: government, healthcare, schools, etc.

These first three things I like to think of as 21st century infrastructure. If you do just these three, you solve half of your decarbonizing problem in decisions that last for decades. You don't need to be mired in day-to-day consumer guilt because you've invested in long-term solutions to your carbon shadow. Succinctly:

1. **Buy, build, or rent electric vehicles to replace your gas guzzlers**

2. **Install heat pumps for home and water heating, and an induction range**

3. **Install solar on your roof, or buy community renewables if you can't.**

As much as we might hope that the future is here and you could just run out and buy those three things, it's not quite as true as we'd like. Let's dive to the next level of detail, where makers will see not only an opportunity to decarbonize their own lives, but also the potential to design and invent new solutions that lower the cost, broaden the appeal, ease the installation, and make the technology more locally relevant.

Stand with the Children

The planet is cooking and it's 2020 already — we're going to have to fix this ourselves. This is essay #2 in a series on how to decarbonize our world. Read the whole series, find hands-on projects to make a difference, and share your ideas at make.co/fix-our-planet.

DR. SAUL GRIFFITH is founder and principal scientist at Otherlab, an independent R&D lab, where he focuses on engineering solutions for a clean energy, net-zero carbon economy. Occasionally makes some pretty cool robots too. Saul got his PhD from MIT, and is a founder or co-founder of makanipower.com, sunfolding.com, voluteinc.com, treau.cool, departmentof.energy, materialcomforts.com, howtoons.com, and more. Saul was named a MacArthur Fellow in 2007.

1. Electrify your ride(s)

Electric cars and plug-in hybrids — You can buy a Tesla (Figure A), Jaguar, BMW, or Bollinger at the top end of town, but right now the more affordable options are Chevy Bolts (Figure B), Fiat 500s, and Hyundai Konas, all of which are fabulous cars.

If you absolutely can't bear the range anxiety, get a plug-in hybrid: try the Prius, the Chevy Volt, or the soccer-van-du-jour, the Chrysler Pacifica. Every year sees many more options arrive on the EV scene, so if what you want doesn't exist, hold out by maintaining your current vehicle until it does. Whatever you do, *don't buy a new car with an internal combustion engine* — that sends the wrong message to everyone.

There are still ample opportunities to be early in electric boats, electric aircraft, and electric RVs to fully decarbonize your transportation fleet, but remember to prioritize electrifying your daily driver, as that's the vehicle that racks up the miles and the carbon.

EV conversions — For the more adventurous, consider electrifying your own classic. Kits for older light cars like the VW Beetle are available, as are services. There are murmurs of a drop-in muscle car "electric small block" motor. Batteries are the piece that's expensive, and motor controls are the bit that's frustrating (and potentially dangerous), but there are many online communities of enthusiasts to join if you want to build an electric hot rod. And there are auction sites where you can pick up a crashed EV and use the parts to do a drivetrain conversion.

Electric bikes, motorcycles, scooters — If you rightly think that roads and car culture are causing harm far beyond merely the climate problem,

then choose electrified public transport or electric bikes (Figure **C**) or motorcycles.

Electrifying a vintage motorcycle is a comparatively easy project if all you want is to toodle around town at 35mph. Electrifying an old moped or a Vespa scooter is a riot, and can be achieved for a few thousand dollars. There's now a profusion of electric bikes on the market, and some great kits for retrofitting an older bike (try converting your own bike on page 64 of this issue).

As noted before, the cost of an electrified vehicle project is dominated by the battery. Commercially they can be obtained for $250–$500 per kWh. Car-sized things will need 300–500Wh per mile of range; motorcycle-sized things 60–100Wh/mile; motor scooter-sized 40–80Wh/mile, and e-bikes 20–40Wh/mile. Choose the range you need, multiply by the battery cost, and roughly double it to estimate the cost of a conversion.

One area where makers could have an impact is in EV control electronics and components. Open source projects like the (quite incredible) VESC (Vedder ESC) motor controller have greatly improved electric skateboards and scooters. Similar projects in battery management systems, displays and UIs, throttles and brakes, and motors could have a big impact on the small vehicle front.

2. Electrify all the heat

Due to the genius of the fossil fuel industry marketing methane as "natural gas," and trumpeting that, per unit of energy, natural gas emits less carbon than coal or oil, the carbon being emitted from our homes and buildings has gone untargeted.

To hit a good climate target under 2°C warming we no longer can wait until we've decommissioned all the coal plants before we decommission the natural gas heating systems that dominate the built environment. People used to say natural gas was a "bridge" fuel to the decarbonized future. Well, we burned all those bridges ... with natural gas.

There are probably three or four things that use natural gas in your home: the furnace (main offender), the stovetop and oven, the water heater, and the clothes dryer. All four of these need to be upgraded, or replaced. Unless you're building a swanky new "Passivhaus" that needs no heating

at all, this looks like a retrofit. Realistically this means heat pumps for your furnace heat and water heater, an electric clothes dryer that's used sparingly, and an induction range for cooking.

Because of the variety and ages of homes, infrastructure, and local laws, decarbonizing our buildings presents a bewildering array of options and corner cases. I'll talk about the "average" path to getting there, but one reason heat appliances are so ripe for maker innovation is that there are many paths. Maybe you live off-grid near a pig farm with methane capture, or a source of wood pellets and walnut shells — you could be doing a local heating fuel economy. By solving a problem in your local community, you're probably coming up with a solution for millions globally.

Hot water heater — The heat pump hot water heater is a drop-in replacement. Rheem and others make them (Figure **D**). It makes a little more noise than traditional heaters, so where it's placed is important.

You can increase its efficiency with more insulation and you could figure out ways to use it to load-shift your house (set it so it only charges when your solar is producing, for example). For the more ambitious: figure out how to lower the fan noise.

Heat pump / furnace / HVAC — The heat pump heating system retrofit is more complicated. Besides the air/ground choice (see "About Heat Pumps" page 20), there are decisions to make: Retrofit the existing forced-air furnace system with all its ductwork? Or replace the entire system with warm water pipes (hydronics) under your floorboards? Hydronic

D

This heat-pump water heater by Rheem is twice as efficient as your old-school water heater, and it's made in the USA.

tesla.com, chevrolet.com, rheem.com

systems are quieter, and give your home better air quality and better comfort. They cost $8–$20 per square foot to install depending on how you do it and whether you do it yourself. This is then connected to your air or ground source heat pump. Hopefully those heat pumps are connected to solar on your roof or community based renewables.

If you can't do hydronic, you can connect the heat pump to a heat exchanger (air handler) so that you can utilize the existing heat ducts.

You might have steam heat or city heat. If city heat, lobby the city to decarbonize the steam source. If it's a boiler in your basement, replace that boiler with a heat pump.

Electric dryer — There's a lot of water in wet clothes. It's hard to get out because all those fibers have enormous surface area. Gas dryers use a huge amount of heat to boil the water off. Electric dryers do the same. They're probably the single biggest load in your house and might even have a special 220V connection.

People are experimenting with ultrasonic and microwave and other kinds of dryers, and you should too — this is a difficult problem to solve. The best solution, of course, is hanging the clothes out to dry. Your clothes will smell and feel better. Maybe a robotic clothesline hanging machine is your next automation project? Minimally, let's program dryers to align their high loads with times of day when solar generation is high.

Electric oven and stove — Cooking with gas is cultural: we've been told it's better. But induction ranges, electric steam ovens, and even hot pots and sous-vide machines are electric options that enable new, more, and better cooking practices. Electronics give us higher fidelity control over the cookware — if you believe computation and controls can improve anything, then obviously this is the path to fancier, better cooking. Get to the kitchen and run some control experiments! Write a cookbook for electric cooking.

With every passing year we learn more about the effects of breathing combusted fossil fuels, and numerous studies now associate exposure to gas cooking with respiratory issues. Electrifying your kitchen will probably improve your health.

About Heat Pumps

A heat pump works by taking a small amount of heat from a large amount of fluid and turning it into a large amount of heat for a smaller amount of fluid. Some describe it as an air conditioner running backwards.

There are two types of heat pump: ***ground-*** or ***earth-source***, and ***air-source***. Heat pumps are most efficient with a low ***lift***, which is the difference between the temperature you want in your house (say, 72°F) and the temperature of your source.

Ground-source pumps use the earth as a reference temperature, which is smart because 4–5 feet below ground everywhere in the continental U.S. it's a pretty constant 55°F–60°F all year round.

Air-source heat pumps use the external air. They're getting much more efficient, but their efficiency drops when the outside air temperature gets really low.

Air source is easier to install, potentially less efficient, and noisier because more fans and fluids. Ground source requires digging or drilling a big heat exchanger into the ground; it's the ultimate system but isn't yet cheap. Invent a robot that can autonomously drill ground loops under houses and you will be richer than Elon Musk.

How do Air-Source Heat Pumps work?

By transferring heat between a house and outside air, these devices trim electricity use by as much as 30 percent to 40 percent in moderate climates.

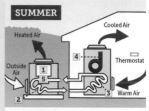

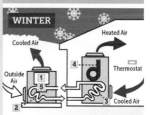

1. Compressor: Increases refrigerant/freon pressure to accept the maximum heat from the air.

2. Condenser: Coils move freon (and with it, hot or cold air) to or from outside air.

3. Evaporator: Coils move freon (and with it, hot or cold air) to or from outside air.

4. Air Handler: Fan blows air into a home's ducts.

5. Reversing Valve: Switches the direction of the freon flow, changing the heat pump's output to hot or cold air (controlled by thermostat).

Photovoltaic solar panels on every home could supply ¼ to ½ of America's total energy needs.

E

3. Put solar on the roof

If every home in America had solar to maximum capacity on its roof, that would supply ¼ to ½ of our total energy supply (Figure **E**). (If we don't cover our roofs we'll have to cover a lot more fragile ecosystems.) If you add in other structures we've already built — buildings, roads, parking lots and garages — we have enough area to produce *more* energy than we need.

It makes sense to produce energy locally to minimize transmission and distribution costs and losses. The average cost of transmission and distribution of electricity is close to 8¢/kWh. Local generation has a huge advantage.

The problem is that solar on roofs in America is expensive. The reason? Bureaucracy. Land of the free, my ass. Somehow Australians and Mexicans and Southeast Asians can install solar for $1–$1.20 per watt, but in the U.S. it's $2.80–$3.20. That's because of liability laws and the cost structures of contracting, and permitting and inspection and union requirements and local building codes   all these "soft costs." The actual modules only cost $0.40 per watt! If we could install them for a fraction more, then everyone would be paying just 3¢–4¢/kWh for electricity.

I think the most noble act of civil disobedience in modern America is to figure out how to install solar on your roof in a way that makes mockery of this situation. Go to it, my friends, innovate not only the installation methods, but also the loopholes that will stop your local officials from preventing your neighbors from decarbonizing!

People complain that modules bolted on the roof look ugly — fix it and design a better solution. There are various efforts in solar tiles, but no one has nailed an easy, robust roofing solution yet — it's a giant market for the maker's taking.

WHAT IF YOU AREN'T A HOMEOWNER?

I get this question a lot. Millennials don't own and think they never will. I still don't know the answer, but you can begin by badgering your landlord, and you can work to change local regulations such that landlords have to prioritize these things.

4. 5. 6. Food, stuff, and services

Indirectly we make plenty more carbon emissions — embedded in the food we eat, the products we buy, and the services we use, including the government that our taxes pay for. But these are harder to quantify. Who made that food or that clothing, how, and using what energy sources?

Nevertheless, there's plenty more you can do. The other decisions in your home:

Insulate and seal — Homes lose a lot of their heat or cool because of air intrusion (gaps under doors and windows) and insufficient insulation of roofs, walls, and windows. Buy, rent, or borrow an infrared camera and take a nighttime walk to find out where the cold or hot air is getting in, and where the insulation can be improved.

Eat better — More than 10% of global emissions are from agriculture, and a good fraction of that is because of our cows, pigs, and sheep. Changing your diet at home is something you can easily do to make a big difference. Becoming vegetarian overnight is culturally and practically unlikely, but eating meat in moderation and not every day is good for your body and for the Amazon.

Repair more — Highly under-appreciated fact: Around 10% of American emissions are carbon embodied in products we import from other countries. Another 20% of our emissions come from industrial production and shipping of our crap here in the U.S. The simplest way to reduce this is to make things last longer. If you can make your computer or bike last twice as long, it'll use half the carbon it might have otherwise.

Makers have fought back on the right to repair. We need to keep fighting back with a culture that appreciates repairing and patching and making our things last longer. The side effect is that we'll be able to own things that are more beautiful that we have a better connection to. ◗

iJessup

YOUTUBER PROFILE
WRITTEN BY JESSIE UYEDA

THE DO-IT-ALL DYNAMO SHARES HER ORIGINS AND APPROACH TO A PROJECT

W hen I was a child, my mother always encouraged me and my three siblings to be creative. In hindsight, I think she was most interested in getting the four of us out of the house. Both of my parents are very creative and capable people and they never worried about us hurting ourselves with power tools or sharp blades because they had already taught us to be mindful of what we were doing and to not be reckless.

So we made stuff. But I never viewed making things as anything other than ancillary to my main objective. I sewed patchwork quilts because my stuffed animals were cold, I cut limbs and twigs off of trees to make bows and arrows because playing Robin Hood required me to do so, and I hand-sawed boards and nailed them together to make stools because I was — and still am — short. In the Uyeda household, these were just things we did.

My background from there is varied in the extreme. I am a classically trained violinist, I was a librarian for 15 years, I was a high school special education teacher, I was production manager for a symphony orchestra, and I was a restaurant manager for 10 years. I never graduated college and I have no training, per se, in any of the things that I do now. I did spend about six months working at a lumber mill. That was one of my favourite jobs.

To me, being a maker means being a problem solver. The question is not "how is this task accomplished?" but, rather, "how do I accomplish this task?" I am not particularly good at any of the things that I do. In fact, I am not very good at most of the things that I do. Being good at something isn't my goal; my goal is to be useful.

When I moved to Boston, it was my understanding that I would be working behind the scenes for my brother Ben's YouTube channel,

Brett McAfee, Jessie Uyeda

If we've learned anything from popular culture, it is that a reluctant protagonist is the best. When Jessie Uyeda agreed to start a YouTube channel at the request of her brother Ben, her reluctance translated into on-screen confidence.

iJessup's videos are typically stark, no-nonsense demonstrations on how to simply get things done. Need a table that looks neat but don't know what you're doing? Here, this is how iJessup would do it. Feel discouraged, like you need to be a big burly guy with a beard or own a whole wood shop to make cool furniture? Just watch some iJessup and get ready for the inspiration to flow.

We connected with Jessie to hear more about her approach to making projects and videos alike. —Make: editors

AT A GLANCE

Channel: youtube.com/ijessup
Year joined: 2015
Projects: Woodworking, blacksmithing, textiles
Subscribers: 33.7k
Total channel views: 513,059
Does own editing: Yes
Cameras: Canon Rebel T6i / iPhone SE / iMovie

HomeMade Modern. I had no desire to make my own videos. In truth, I have stage fright and feel very awkward on camera... also, in person. So, when Ben asked me what I wanted to call my channel, I told him that I didn't want to have my own channel. His response was, "You have to." My response? "Okay."

I decided that my angle would be to show people that the biggest thing that can keep them from making things themselves is the idea that they can't. Again, my biggest goal in life is to be a useful person, no matter what my experience level. I'd love to know that others can find that out about themselves, as well. My first video is one of the most basic things anyone could ever make: a coffee table with 2"×2" balusters for legs and a floor tile for the top. I had no idea what I was doing so the simplicity of the project lent itself well to my "skill" set.

With my videos, I do everything myself. There is a definite benefit to living with two other content creators in that there are always people to bounce ideas off of and get input on what would be a good title for a video or which photo would make the best thumbnail but we all do our own filming and editing.

The most difficult part of the process for me is, by far, the initial idea. I never know what to make. I get really caught up in wanting to build something that is functional and something that I would potentially use or need. But, there really isn't much that I need. So, then the problem is: what do I have the skill set to build and how do I make it interesting? I feel like most people in my field are teeming with ideas and that's where I have the most trouble. I'm not good at drawing and I don't know how to 3D model so I often just skip the planning phase and go straight to building so I can figure out the problems as they arrive.

I shoot my YouTube videos and thumbnails on a Canon Rebel T6i and my instagram is mostly shot on my iPhone SE. All of my videos are edited using iMovie. I don't know anything about cameras or editing so I just do what makes sense to me with what I already have. People are most interested in the content ,and as long as it doesn't look terrible it doesn't really matter what your equipment is.

Brett McAfee, Jessie Uyeda

PEOPLE ARE MOST INTERESTED IN THE CONTENT, AND AS LONG AS IT DOESN'T LOOK TERRIBLE IT DOESN'T REALLY MATTER WHAT YOUR EQUIPMENT IS.

I am now embarking on my biggest project ever: an entire house. I spent last year trying to talk my brother into buying me a small house to fix up. What I got was a long abandoned, hoarder's house that has been home to several squatters and legions of rats over the years. It is both exhilarating and terrifying. While I am not building an entire house from the ground up, I am almost certainly stripping the existing house down to the studs and taking it from there. There is so much that I will be learning from this huge project and I am so very excited to capture all my triumphs and failures and share them with anyone who has an interest in doing things themselves. ●

The Rebble Alliance

With server shutdowns looming, a group of Pebble enthusiasts battled to keep their beloved smartwatches operational

Written by Kevin Purdy

Kevin Purdy is a writer for iFixit, an online repair guide and store for parts and tools. He lives in Buffalo, New York.

In the early summer of 2018, you could buy an Apple Watch with built-in GPS, wireless payments, and speakers that buzz water out after a swim. Meanwhile, Katharine Berry was hustling to keep five-year-old watches with black-and-white screens alive. Berry had worked for Pebble, maker of the first notable smartwatch. The company was acquired by Fitbit and shut down a year and a half earlier. Now Fitbit was turning off the servers that fed Pebble's apps, weather, and other useful data.

But Berry and a cadre of crafty enthusiasts, the Rebble Alliance, had prepared for this moment. They had archived Pebble's web and development assets, opened up the devices' firmware a bit, and worked with former Pebble and Fitbit developers inside a Discord channel. Berry, between jobs, sprinted for two weeks to code a replacement cloud infrastructure. She guessed that, if they could pull it off, maybe a thousand people, at most, would try it out.

One morning seven months later, Berry realized that Rebble had 100,000 accounts. Today, more than 212,000 accounts have been created — more than 10% of the two million Pebbles ever sold — and nearly 9,000 have subscribed. Press coverage certainly helped. But really it is the Rebblers' enthusiasm that keeps their watches running, precisely because Pebbles are *not* modern, in all the best ways.

Rebble is an inspiring repair story, and the way Pebble enabled this second life is a path that every gadget manufacturer should strive to emulate. Pebble created an open (and open-source) environment for developers and enthusiasts. As a direct result, Rebble is saving thousands of gadgets from the bin and building a real community around dogged longevity. Keeping Pebbles running, in the face of much fancier options, knitted the community together.

Arduino Roots

Prior to founding Pebble, Eric Migicovsky was studying abroad at the University of Delft, Netherlands in 2008. He fit in with his sturdy, reliable cruiser bike (known to locals as an *opafiets,* or "Grandpa bike"), but he knew he would crash if he kept checking his text messages while riding. In his dorm room,

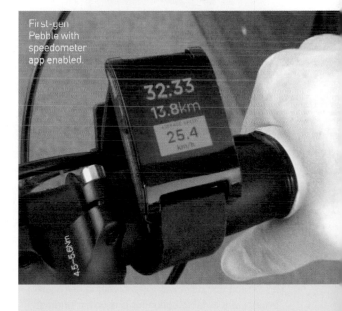

First-gen Pebble with speedometer app enabled.

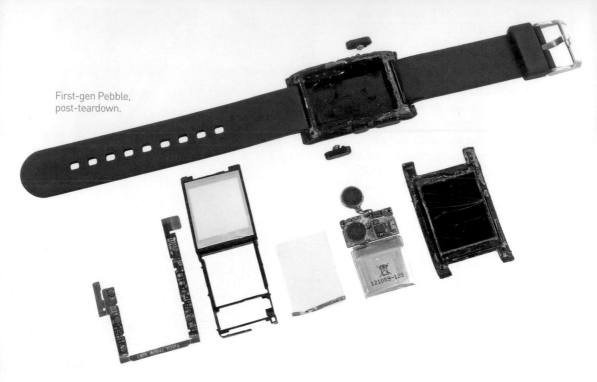

First-gen Pebble,
post-teardown.

he patched together an Arduino controller, a few buttons, a battery, and the screen from a disassembled Nokia 3310. His first idea was a bike computer, but "someone was like, you should probably just make it a watch," he later said.

Migicovsky's refined model, called inPulse and built for BlackBerry phones, got him into startup bootcamp Y Combinator in 2011. He impressed founder Paul Graham, who said Migicovsky was the most likely pick to be "the next Steve Jobs."

Creating the next watch required a lot more money, and venture capitalists are notoriously cautious about hardware. So in the spring of 2012, Migicovsky turned to Kickstarter. Yet again, his timing was keen. Kickstarter projects were becoming not just a way to fund your cousin's art project, but a viable option for pre-order funding. Months earlier, a video game and iPhone dock had broken the $1 million mark on the same day.

The launch of Migicovsky's next watch, called Pebble, left Kickstarter's records in the dust, raising $1 million in 28 hours and finishing with a new milestone for Kickstarter at the time — over $10 million.

First Mover Disadvantage

The first Pebble, shipped after delays in January 2013, was a plastic watch with a black-and-white e-paper screen. Between that and the 2014 Steel release, Pebble had proven that people appreciated an alternative to constant phone-opening. But then came the inevitable: Apple debuted its Watch. Could Pebble perfect its device and broaden its appeal before Apple ate the entire field?

The Kickstarter for the next Pebble, the Time, launched exactly two months before the Apple Watch would ship. The campaign still holds the record for the largest-ever Kickstarter at over $20 million. The Time, and its fancier Time Steel version, added a first-of-its-kind 64-color e-paper screen, voice dictation, and a novel, whimsically animated OS.

At the same time Pebble was breaking records and touting its indie appeal, the company had stopped being profitable. One company source told *Business Insider* that Apple had "sucked up all the oxygen." Pebble missed its sales goals in late 2015, and its Black Friday sales in 2015 were down from the year before. What followed were layoffs, a failed acquisition by Intel, and trouble finding more capital. Pebble's final Kickstarter in May 2016 was, in effect, a bridge loan from its fans.

The next wave of Pebbles focused on fitness, something Apple was already pivoting toward with its second Watch, but Pebble didn't have Apple's money. After months of last-ditch fundraising attempts, Fitbit acquired Pebble's software assets and engineer hiring rights in December 2016 for a scant $23 million. While Fitbit would not officially support Pebble's customers, Migicovsky worked out a deal that would refund Kickstarter pre-orders, and, he hoped, keep the more than two million Pebbles sold, and their apps, working for as long as possible.

In hindsight, he nailed it. Pebble has been the most successful hardware company failure in history. Compare this to Revolv, whose acquisition by Nest led to an abrupt shutoff of smart homes around the world, or personal cloud device Lima, or, really, any Android device more than a couple years old.

3D-printed Pebble 2 cases, alongside a Pebble 2 display.

The last watch to ship, the Pebble 2.

The Panic Store

News of the shutdown shook the Pebble community. The official message was that Pebbles "will work normally for now," but "Functionality or service quality may be reduced down the road." Without web services from Pebble, watches would lose their app store, language packs, voice dictation, weather, and even the icons on their notifications. Nobody knew anything for sure, other than that Pebble, the company, no longer existed. "We were all just freaking out," says IShotJr, a longtime Pebble developer, hardware hacker, and community organizer. "We're all just sitting on the Discord, panicking."

From the rubble formed Rebble, a team of motivated fans, developers, and ex-employees, rushing to reproduce years of development in a matter of days. Frantic to document critical APIs and development tools before the servers shut off, they grabbed everything they could. The first replacement app store appeared quickly, aptly code-named Panic App Store. Within a few days, they had firmware, core and third-party apps, all the dev tools, and more. And they preserved it all on a wiki, right down to the pinouts.

> # Pebble has been the most successful hardware company failure in history.

The Rebble App Store, a complete replica of the Pebble App Store, as it looks today.

Pebble Time,
under disassembly.

Meanwhile, Fitbit kept the servers running longer than expected. But the axe would fall in June 2018, and a replacement was needed. Using her inside knowledge of Pebble's server setup, and painstakingly working through a man-in-the-middle proxy, Berry, the ex-Pebbler, created replacement web services for nearly everything Pebble had provided. With just 16 days until the shutdown, Rebble opened up account sign-ups. When Fitbit finally killed Pebble's servers, Rebble was ready the next day.

More than 177,000 people have connected their devices to Rebble's services. Not everything can be free, because the APIs for voice dictation and weather are not cheap — $750,000 per year, if 100,000 people used them, Berry said. And yet, nearly 9,000 people pay yearly subscriptions.

Keeping Pebbles Ticking

Pebble watches were built to hit low price points, with most models selling below $150. They are small devices, made by a company without a lot of manufacturing leverage. Most started out with week-long battery life, but the oldest devices are nearly seven years old now. Pebbles are prone to certain mechanical failures, and despite the success of the Rebble community, some are quite

tricky to repair. But Rebble is working on it. The iFixit community has provided some repair guides, and is sourcing batteries and spare parts.

Two Rebblers, Astosia and Tation, have opened a Shapeways store full of 3D-printed Pebble 2 cases and buttons for sale, because the silicone buttons on the original Pebble 2 are breaking down over time. "Removing the screen is stress-inducing," Astosia said, but other than that, she claims a Pebble 2 case swap is a fairly straightforward transplant of internals. Pebble itself released the 3D printing files for all its watches, so fans have been experimenting.

All Pebbles will eventually die, though. The goal, then, is to create an entirely self-built, reverse-engineered version of the Pebble's original firmware: RebbleOS (github.com/pebble-dev/RebbleOS). And then find, or maybe hack together, future watch hardware on which it might run. This idea isn't as crazy as it sounds — Pebble is based on FreeRTOS, an open-source kernel that supports a vast array of hardware. Already, some proofs of concept are running on Rebblers' workspaces.

Joshua Wise, a lead developer with infectious enthusiasm, thinks a completely Rebble-built watch is more of a project management challenge

iFixit, Pebble

Pebble 2, torn down.

than a moonshot. The biggest challenge to keeping Rebble running on existing Pebbles is triaging Pebble's smartphone apps, which occasionally disappear from their respective iOS and Android stores. Beyond that, it's about freeing up time to experiment and dream of the future — and getting Bluetooth to work reliably, which is "always a pain in the butt."

Migicovsky, now a partner at Y Combinator, is actively using Rebble, and is proud of Pebble's longevity. "Every so often I take out my OG Pebble that I grabbed off the assembly line — one of the first ones," Migicovsky says. "Its manufacture date was December 26, 2012!"

Not many modern, web-connected devices live on for years after their maker goes out of business and shuts down its servers. Fewer still have not only an active repair and support community, but a forward-looking mission. Rebble is a welcoming, open-source, community-minded effort, with a responsible financial model behind it. It's hard to believe it exists, and feels like some still-raw chunk of 2013 tech optimism that can't possibly survive into the future.

Except, it might. ◗

More than 177,000 people have connected their devices to Rebble's services.

Pebble's Timeline interface, focused on calendar events and reminders.

This article originally ran on iFixit's website. You can read it here: ifixit.com/News/rebble-with-a-cause-how-pebble-watches-got-their-amazing-afterlife

Marvel of Miscellany

Written by Caleb Kraft

BEST DEAL

Subscribe today and get Make: delivered right to your door for just $39.99! You **SAVE 33% OFF** the cover price!

Name _____ (please print) _____

Address/Apt. _____

City/State/Zip _____

Country _____

Email (required for order confirmation) _____

☑ **One Year**
$39.99

BOANS3

☐ **Payment Enclosed**   ☐ **Bill Me Later**

Make:

Make: currently publishes 4 issues annually. Allow 4-6 weeks for delivery of your first issue. For Canada, add $9 US funds only. For orders outside the US and Canada, add $15 US funds only.

**To order go to:
Makezine.com/save33**

BUSINESS REPLY MAIL
FIRST-CLASS MAIL   PERMIT NO. 865   NORTH HOLLYWOOD, CA

POSTAGE WILL BE PAID BY ADDRESSEE

Make:

PO BOX 17046
NORTH HOLLYWOOD CA 91615-9186

City Museum

St. Louis' **CITY MUSEUM** is a hands-on, DIY wonderland for everyone

I f you were to give a 100-year-old, 10-story, abandoned shoe factory to Dr. Seuss and M.C. Escher, and tell them to make the world's craziest jungle gym, you'd end up with the City Museum in St. Louis, Missouri.

The name "museum" may mislead a few people. This isn't your typical meander and browse sort of affair. Instead, what you'll find are the preserved remnants of demolished buildings, cobbled together into works of art that you can climb over, crawl through, squeeze past, and admire in awe.

The City Museum was created by an artist named Bob Cassilly. Together with his wife and a group of hand-picked artisans he started collecting architectural elements and industrial extras from St. Louis construction sites, and upcycling them into art. Every year there is a building season in the middle of winter where the museum expands further. Even though Cassilly passed away a few years ago, the workers, who call themselves the Cassilly Crew, work hard to keep the dream alive and keep the museum growing.

Caleb Kraft is senior editor for *Make:* magazine. He discovered he was claustrophobic while crawling through a sub-floor tunnel, carrying a 2-year-old, with a line of people behind him, at the City Museum in St. Louis.

To say you can "touch" the art would be a major understatement. Anywhere you stand at the City Museum, you'll find activity all around you. Everything is sculptural art, and has people in it or on it. You may crawl through the gaping maw of a whale sculpture to find yourself in a mystical labyrinth of simulated caves, only to exit through a tunnel along the ceiling into the floor above where a circus act is in mid-swing. Nothing here is built to any kind of a typical building code (you have to sign a waiver to get in); people sometimes get stuck in narrow passages or can't pull themselves up multiple stories through coiled steel, but then they just turn around and go back and everything is fine.

Inside, the structures and art are mainly composed of smaller items; baking tins line one wall to give it texture, and an old bank safe serves as a tunnel from one area to another. Outside, however, stands what the crew says is the largest outdoor sculpture in the world. You can climb through rebar tunnels — suspended stories in the air — to go from an airplane to a fire truck and even into a school bus perched perilously over the edge of the building. If you're cold you can sit around the bonfire while watching people play in the massive ball pit.

"In a way, it is similar to a regular museum," says one Cassilly Crew member. "These things attached to the walls are the processes of manufacturing from days past. As new technology comes in, the old stuff gets displaced and finds a home here at the City Museum."

If you explore long enough, you may find that there is actually a small section that has items laid out more like a traditional gallery. This museum

City Museum, Caleb Kraft

> "The point is not to learn every fact, but to say, 'Wow, that's wonderful.' And if it's wonderful, it's worth preserving." —*Bob Cassilly*

within the museum carries a historical selection of things like gargoyles, decorative concrete blocks, and doorknobs. After sliding down the 10-story-tall spiral shoe chute, a quiet stroll through history can feel like a welcome recovery.

It is hard to explain the vastness of the City Museum. I've visited multiple times and on each expedition I was surprised with areas I'd never seen before. A crack in the wall that I may have overlooked multiple times could lead you to an entirely unexplored branch, and that kind of perpetual awe and delight feels priceless.

The crew is still expanding the art to this day, slowly filling every nook and cranny of the 600,000-square-foot building. Cassilly's memory not only lives on, but is thriving in the experiences children and adults alike have when they visit. I hope it keeps growing and changing forever. ⊘

The Cost of Convenience

WRITTEN BY MIKE SENESE

"The surveillance capitalists are driving a dishonest bargain with the clientele, while the Internet per se is dominated by computer-crime mafia and intelligence services now; it's genuinely hazardous."

—Bruce Sterling

Over the past decade, the space between us humans and our gadgets has become almost imperceptible. Our phones, computers, and even our doorbells act as an extension to our eyes and ears. An entirely new class of consumer device, the voice assistant, has become ubiquitous. Our cars are even able to drive themselves for us. We now live in the future.

There's a part of this future that we didn't daydream about, though — that our technology would transmit all the data it observes to corporate servers, which log our daily locations, habits, and interests. With this, the companies are now targeting us for individualized marketing. We've sold off our privacy and we don't even know it.

"Surveillance Capitalism describes the current economic model of technology companies that make revenue by surveilling our online lives, gathering data that is processed and transformed to result in targeted advertising packages," write Andrea Krajewski and Max Krüger in the opening to *The State of Responsible IoT 2019*, published as part of the annual ThingsCon gathering (thingscon. org). "The better the data, the more likely it is that we do what is expected of us: buy what we are shown. Surveillance Capitalism is therefore not only an economic model, it is a form of control over our behavior."

This stockpiling of data also becomes a target for more (and worse) privacy breaches in the future. "Best case, there will be a large data leak that exposes people's most personal information about their homes, habits, and associations," says Gennie Gebhart, associate director of research for the Electronic Frontier Foundation (EFF). "In the worst case, bad actors from law enforcement to domestic abusers will use these home surveillance systems against people in even more insidious ways than we're seeing now." To combat this, the EFF has created Surveillance Self-Defense (ssd.eff.org), a resource for protecting individuals' online privacy and security.

We've become more aware of the tradeoffs these devices incur, yet many do want the conveniences they offer. There are, however, no commercial options that really protect you — you can't pay to keep free of the surveillance. You've got to DIY it. Read on and we'll show you how. ⊘

Mark Madeo

Open Source

WRITTEN BY KATHY REID

Voice
for Makers

Command your own voice assistant without the Big Tech eavesdroppers and data snoops, using free and private tools

KATHY REID is a master's student in applied cybernetics at Australian National University. She was previously director of developer relations at Mycroft AI.

Science fiction has whetted our imagination for helpful voice assistants. Whether it's JARVIS from *Iron Man*, KITT from *Knight Rider*, or Computer from *Star Trek*, many of us harbor a desire for a voice assistant to manage the minutiae of our daily lives. Speech recognition and voice technologies have advanced rapidly in recent years, particularly with the adoption of Siri, Alexa, and Google Home.

However, many in the maker community are concerned — rightly — about the privacy implications of using commercial solutions. Just how much data do you give away every time you speak with a proprietary voice assistant? Just what are they storing in the cloud? What free, private, and open source options are available? Is it possible to have a voice stack that doesn't share data across the internet?

Yes, it is. In this article, I'll walk you through the options.

WHAT'S IN A VOICE STACK?

Some voice assistants offer a whole *stack* of software, but you may prefer to pick and choose which *layers* to use.

» **WAKE WORD SPOTTER** — This layer is constantly listening until it hears the *wake word* or *hot word*, at which point it will activate the speech-to-text layer. "Alexa," "Jarvis," and "OK Google" are wake words you may know.

» **SPEECH TO TEXT (STT)** — Also called *automatic speech recognition (ASR)*. Once activated by the wake word, the job of the STT layer is just that: to recognize what you're saying and turn it into written form. Your spoken phrase is called an *utterance*.

» **INTENT PARSER** — Also called *natural language processing (NLP)* or *natural language understanding (NLU)*. The job of this layer is to take the text from STT and determine what action you would like to take. It often does this by recognizing *entities* — such as a time, date, or object — in the utterance.

» **SKILL** — Once the intent parser has determined what you'd like to do, an application or handler is triggered. This is usually called a *skill* or *application*. The computer may also create a reply in human-readable language, using *natural language generation (NLG)*.

» **TEXT TO SPEECH** — Once the skill has completed its task, the voice assistant may acknowledge or respond using a synthesized voice.

Some layers work *on device*, meaning they don't need an internet connection. These are a good option for those concerned about privacy, because they don't share your data across the internet. Others do require an internet connection because they offload processing to cloud servers; these can be more of a privacy risk.

Before you pick a voice stack for your project you'll need to ask key questions such as:

- What's the interface of the software like — how easy is it to install and configure, and what support is available?
- What sort of assurances do you have around the software? How accurate is it? Does it recognize your accent well? Is it well tested? Does it make the right decisions about your intended actions?
- What sort of context, or use case, do you have? Do you want your data going across the internet or being stored on cloud servers? Is your hardware constrained in terms of memory or CPU? Do you need to support languages other than English?

All-in-One DIY Voice Assistants

Tool	Best for...	Best installed on...	Best if you know a bit about...	Not so good if...
Jasper	Beginners	Raspberry Pi	Python	You need support
Rhasspy	Experienced developers	Your own machine	Docker	You need support
Mycroft	Experienced developers	Raspberry Pi or a Linux machine	Python, bash	You don't want any internet connection
Almond	Beginners	Linux under Gnome, or use web interface	Javascript	You want to run on RasPi, or you need support

ALL-IN-ONE VOICE SOLUTIONS

If you're looking for an easy option to start with, you might want to try an all-in-one voice solution. These products often package other software together in a way that's easy to install. They'll get your DIY voice project up and running the fastest.

Jasper (jasperproject.github.io) is designed from the ground up for makers, and is intended to run on a Raspberry Pi. It's a great first step for integrating voice into your projects. With Jasper, you choose which software components you want to use, and write your own skills, and it's possible to configure it so that it doesn't need an internet connection to function (Figure **A**).

Rhasspy (rhasspy.readthedocs.io) also uses a modular framework and can be run without an internet connection. It's designed to run under Docker and has integrations for NodeRED and for Home Assistant, a popular open source home automation software (Figure **B**).

Mycroft (mycroft.ai) is modular too, but by default it requires an internet connection. Skills in Mycroft are easy to develop and are written in Python 3; existing skills include integrations with Home Assistant and Mozilla WebThings. Mycroft also builds open-source hardware voice assistants (Figure **C**) similar to Amazon Echo and Google

Home. And it has a distribution called Picroft specifically for the Raspberry Pi 3B and above (build your own Picroft on page 58 of this issue).

Almond (almond.stanford.edu) is a privacy-preserving voice assistant from Stanford that's available as a web app, for Android (Figure **D**), or for the GNOME Linux desktop. Almond is very new on the scene, but already has an integration with Home Assistant. It also has options that allow it to run on the command line, so it could be installed on a Raspberry Pi (with some effort).

The languages supported by all-in-one voice solutions are dependent on what software options are selected, but by default they use English. Other languages require additional configuration.

WAKE WORD SPOTTERS

PocketSphinx (github.com/cmusphinx/pocketsphinx) is a great option for wake word spotting. It's available for Linux, Mac, Windows platforms, as well as Android and iOS; however, installation can be involved. PocketSphinx works on-device, by recognizing *phonemes*, which are the smallest units of sound that make up a word.

For example, *hello* and *world* each have four phonemes:

hello H EH L OW
world W ER L D

The downside of PocketSphinx is that its core

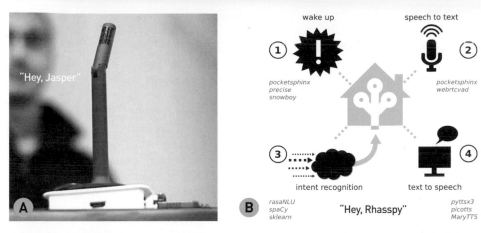

A "Hey, Jasper"

B "Hey, Rhasspy"

wake up
1
pocketsphinx
precise
snowboy

speech to text
2
pocketsphinx
webrtcvad

intent recognition
3
rasaNLU
spaCy
sklearn

text to speech
4
pyttsx3
picotts
MaryTTS

C "Hey, Mycroft"

D "Hey, Almond"

developers appear to have moved on to a for-profit company, so it's not clear how long PocketSphinx or its parent CMU Sphinx will be around.

Precise by Mycroft.AI (github.com/MycroftAI/mycroft-precise) uses a recurrent neural network to learn what are and are not wake words. You can train your own wake words with Precise, but it does take a lot of training to get accurate results.

Snowboy (github.com/Kitt-AI/snowboy) is free for makers to train your own wake word, using Kitt.AI's (proprietary) training, but also comes with several pre-trained models, and wrappers for several programming languages including Python and Go. Once you've got your trained wake word, you no longer need an internet connection. It's an easier

option for beginners than Precise or PocketSphinx, and has a very small CPU footprint, which makes it ideal for embedded electronics. Kitt.AI was acquired by Chinese giant Baidu in 2017, although to date it appears to remain as its own entity.

Porcupine from Picovoice (github.com/Picovoice/porcupine) is designed specifically for embedded applications. It comes in two variants: a complete model with higher accuracy, and a compressed model with slightly lower accuracy but a much smaller CPU and memory footprint. It provides examples for integration with several common programming languages. Ada, the voice assistant recently released by Home Assistant, uses Porcupine under the hood (github.com/home-assistant/ada).

SPEECH TO TEXT

Kaldi (kaldi-asr.org) has for years been the go-to open source speech-to-text engine. Models are available for several languages, including Mandarin. It works on-device but is notoriously difficult to set up, not recommended for beginners. You can use Kaldi to train your own speech-to-text model, if you have spoken phrases and recordings, for example in another language. Researchers in the Australian Centre for the Dynamics of Language have recently developed **Elpis** (github.com/coedl/elpis), a wrapper for Kaldi that makes transcription to text a lot easier. It's aimed at linguists who need to transcribe lots of recordings.

CMU Sphinx (cmusphinx.github.io), like its child PocketSphinx, is based on phoneme recognition, works on-device, and is complex for beginners.

DeepSpeech (github.com/mozilla/deepspeech), part of Mozilla's Common Voice project (voice.mozilla.org), is another major player in the open source space that's been gaining momentum. DeepSpeech comes with a pre-trained English model but can be trained on other data sets — this requires a compatible GPU. Trained models can be exported using TensorFlow Lite for inference, and it's been tested on an RasPi 4, where it comfortably performs real-time transcriptions. Again, it's complex for beginners.

INTENT PARSING AND ENTITY RECOGNITION

There are two general approaches to intent parsing and entity recognition: *neural networks* and *slot matching*. The neural network is trained on a set of phrases, and can usually match an utterance that "sounds like" an intent that should trigger an action. In the slot matching approach, your utterance needs to closely match a set of predefined "slots," such as "play the song [*songname*] using [*streaming service*]." If you say "play Blur," the utterance won't match the intent.

Padatious (mycroft-ai.gitbook.io/docs/mycroft-technologies/padatious) is Mycroft's new intent parser, which uses a neural network. They also developed **Adapt** (github.com/mycroftai/adapt) which uses the slot matching approach.

For those who use Python and want to dig a little deeper into the structure of language, the **Natural Language Toolkit** (nltk.org) is a powerful tool, and can do "parts of speech" tagging — for example recognizing the names of places.

Rasa (github.com/rasahq/rasa) is a set of tools for conversational applications, such as chatbots, and includes a robust intent parser. Rasa makes predictions about intent based on the entire context of a conversation. Rasa also has a training tool called Rasa X, which helps you train the conversational agent to your particular context. Rasa X comes in both an open source community edition and a licensed enterprise edition (rasa.com/product/pricing).

Picovoice also has **Rhino** (github.com/picovoice/rhino), which comes with pre-trained intent parsing models for free. However, customization of models — for specific contexts like medical or industrial applications — requires a commercial license.

TEXT TO SPEECH

Just like speech-to-text models need to be "trained" for a particular language or dialect, so too do text-to-speech models. However, text to speech is usually trained on a single voice, such as "British Male" or "American Female."

eSpeak (github.com/espeak-ng) is perhaps the best-known open source text-to-speech engine. It supports over 100 languages and accents, although the quality of the voice varies between languages. eSpeak supports the Speech Synthesis Markup Language format, which can be used to add inflection and emphasis to spoken language. It is available for Linux, Windows, Mac, and Android systems, and it works on-device, so it can be used without an internet connection, making it ideal for maker projects.

Festival (festvox.org/festival) is now quite dated, and needs to be compiled from source for Linux, but does have around 15 American English voices available. It works on-device. It's mentioned here out of respect; for over a decade it was considered the premier open source text-to-speech engine.

Mimic2 (github.com/mycroftai/mimic2) is a **Tacotron** (google.github.io/tacotron) fork from Mycroft AI, who have also released the **Mimic Recording Studio** to allow you to build your own text-to-speech voices. To get a high-quality voice requires up to 100 hours of "clean" speech, and Mimic2 is too large to work on-device, so you need to host it on your own server or connect your device to the Mycroft Mimic2 server. Currently it only has a pre-trained voice for American English.

Mycroft's earlier **Mimic TTS** (github.com/mycroftai/mimic2) can work on-device, even on a Raspberry Pi, and is another good candidate for maker projects. It's a fork of CMU Flite.

Mary Text to Speech (mary.dfki.de) supports several, mainly European languages, and has tools for synthesizing new voices. It runs on Java, so can be complex to install.

So, that's a map of the current landscape in open source voice assistants and software layers. You can compare all these layers in the chart on pages 44–45. Whatever your voice project, you're likely to find something here that will do the job well — and will keep your voice and your data private from Big Tech.

WHAT'S NEXT FOR OPEN SOURCE VOICE?

As machine learning and natural language processing continue to advance rapidly, we've seen the decline of the major open source voice tools. CMU Sphinx, Festival, and eSpeak have become outdated as their supporters have adopted other tools, or maintainers have gone into private industry and startups.

We're going to see more software that's free for personal use but requires a commercial license for enterprise, as Rasa and Picovoice do today. And it's understandable; dealing with voice in an era of machine learning is data intensive, a poor fit for the open source model of volunteer development. Instead, companies are driven to commercialize by monetizing a centralized "platform as a service."

Another trajectory this might take is some form of value exchange. Training all those neural networks and machine learning models — for STT, intent parsing, and TTS — takes vast volumes of data. More companies may provide software on an open source basis and in return ask users to donate voice samples to improve the data sets. Mozilla's Common Voice follows this model.

Another trend is voice moving on-device. The newer, machine-learning-driven speech tools originally were too computationally intensive to run on low-end hardware like the Raspberry Pi. But with DeepSpeech now running on a RasPi 4, it's only a matter of time before the newer TTS tools can too.

We're also seeing a stronger focus on personalization, with the ability to customize both speech-to-text and text-to-speech software.

WHAT WE STILL NEED

What's lacking across all these open source tools are user-friendly interfaces to capture recordings and train models. Open source products must continue to improve their UIs to attract both developer and user communities; failure to do so will see more widespread adoption of proprietary and "freemium" tools.

As always in emerging technologies, standards remain elusive. For example, skills have to be rewritten for different voice assistants. Device manufacturers, particularly for smart home appliances, won't want to develop and maintain integrations for multiple assistants; much of this will fall to an already-stretched open source community until mechanisms for interoperability are found. Mozilla's WebThings ecosystem (see page 50) may plug the interoperability gap if it can garner enough developer support.

Regardless, the burden rests with the open source community to find ways to connect to proprietary systems (see page 46 for a fun example) because there's no incentive for manufacturers to do the converse.

The future of open source rests in your hands! Experiment and provide feedback, issues, pull requests, data, ideas, and bugs. With your help, open source can continue to have a strong voice. ○

Options for a DIY Open Source Voice Stack

	Software	Comments	Open source software license
WAKE WORD SPOTTERS	**PocketSphinx**	Works on-device, low memory footprint. Available on several platforms. Support via SourceForge forum. Available only for English, long-term feasibility of project questionable.	Uses its own specific license, very similar to the BSD license
	Precise	Can be used to train custom wake words. Free for personal and commercial use. Only available for Linux. Has multiple dependencies and complex setup process.	Apache 2 license
	Snowboy	Easy to train; has wrappers for several languages. Very small CPU footprint. Free for personal use; commercial use requires a commercial license.	Apache license
	Porcupine	Very small CPU and memory footprint, wrappers for several languages. Several development language examples given. Only free for personal use, commercial use requires a commercial license.	Apache 2 license
SPEECH TO TEXT	**Kaldi**	Mature software product, available for multiple languages. Able to train on own language data. Very difficult to install and configure; steep learning curve.	Apache 2 license
	DeepSpeech	Backed strongly by Mozilla, will run on-device on an RasPi 4 using pre-trained models or after training. Bleeding edge; only runs on Linux, has significant dependencies for installation. Word Error Rate (WER) of 7.5%	Mozilla Public License 2.0
	CMU Sphinx	Mature software product, models available in several languages. Difficult to install; long-term feasiblity of project questionable.	BSD
INTENT PARSING	**Adapt**	Free for both personal and commercial use. Uses slot-matching approach.	Apache 2 license
	Padatious	Free for both personal and commercial use. Uses neural network approach.	Apache 2 license
	NLTK	Robust linguistic software for analysis, provides advanced parts of speech tagging. Available only in Python.	Apache 2 license
	Rasa	Provides intent parsing based on conversational context. Well documented, easy to get started with. Customization of context requires a commercial license.	Apache 2 license
	Rhino	Well documented, easy to get started with. Customization of context requires a commercial license.	Apache 2 license
TEXT TO SPEECH	**eSpeak**	Robust software, support for over 100 languages. Works on device. Quality of voice varies.	GPL
	Festival	Works on device, but is dated.	X11-type license, unrestricted commercial and non-commercial use
	Mimic2	Has a range of tools to train your own voice. Too computationally intensive to work on device.	Apache 2 license
	Mimic	Works on device, but has limited range of available voices.	Apache 2 license
	Mary TTS	Range of languages supported. Runs on Java, complex to install.	LGPL

On device or requires connectivity?	Auspicing organization	Support available?
On device	Carnegie Mellon University	Community-supported forum (sourceforge.net/p/cmusphinx/discussion)
On device	Mycroft AI (a for-profit company)	Company-backed forum and online community (community.mycroft.ai/search?q=precise and chat.mycroft.ai/community/channels/wake-word)
On device once trained, requires internet access to Kitt.AI platform to train	Kitt.AI (a for-profit company owned by Baidu)	Community-supported forum for those without a commercial license (groups.google.com/a/kitt.ai/forum/#!forum/snowboy-discussion)
On device. Customized training requires internet access to Picovoice.AI platform	Picovoice.AI (a for-profit company)	No support forum, but GitHub issues are available to browse (github.com/Picovoice/porcupine)
On device	Johns Hopkins University, with support from many other research bodies	Community-supported forum (sourceforge.net/p/kaldi/discussion)
On device, but only for capable hardware (RasPi 4 minimum)	Mozilla (a for-profit company)	Community-backed forum (discourse.mozilla.org/c/deep-speech)
On device	Carnegie Mellon University	Community-backed forum (sourceforge.net/p/cmusphinx/discussion)
On device	Mycroft AI (a for-profit company)	Company-backed forum and online community (community.mycroft.ai/search?q=adapt and chat.mycroft.ai/community/channels/adapt)
On device	Mycroft AI (a for-profit company)	Company-backed forum and online community (community.mycroft.ai/search?q=padatious and chat.mycroft.ai/community/channels/padatious)
On device	Not auspiced by an organization	No support forum, but GitHub issues are available to browse (github.com/nltk/nltk/issues)
On device after initial training; custom training requires internet connectivity	Rasa (a for-profit company)	Company-backed forum (forum.rasa.com)
On device after initial training; custom training requires internet connectivity	Picovoice (a for-profit company)	No support forum, but GitHub issues are available to browse (github.com/Picovoice/rhino/issues)
On device	Not auspiced by an organization	Community-backed forum (sourceforge.net/p/espeak/discussion)
On device	University of Edinburgh	Forum and mailing lists currently offline
No, but can be hosted by user	Mycroft AI (a for-profit company)	Company-backed forum and online community (community.mycroft.ai/search?q=mimic2 and chat.mycroft.ai/community/channels/mimic)
On device	Mycroft AI (a for-profit company)	Company-backed forum and online community (community.mycroft.ai/search?q=mimic and chat.mycroft.ai/community/channels/mimic)
On device	DFKI (a German for-profit company)	Mailing list (www.dfki.de/mailman/listinfo/mary-users)

Paralyzing Privacy Parasite!

Plop it on Alexa or Google assistants to jam their eavesdropping and confuse their data, but still command all their functions

WRITTEN BY TORE KNUDSEN AND BJØRN KARMANN

TIME REQUIRED:
1–2 Hours plus 3D printing

DIFFICULTY:
Intermediate

COST:
$70–$100

MATERIALS
- » **Raspberry Pi 3 Model A+ single-board computer** with 5V power supply and microSD card. (A RaspPi B model is too big to fit.)
- » **ReSpeaker 2-Mics Pi Hat microphone board** seeedstudio.com
- » **Miniature speakers, 16mm, 50Ω (2)** such as Visaton # 2816, Amazon #B003EA3VPO
- » **Wood screws, 2mm×10mm (4)**
- » **Hookup wire**
- » **JST 2.0 connector or 3.5mm stereo cable**
- » **3D printed Alias housing** Design inspired by parasitic fungus! Download the free STL files for printing at instructables.com/id/Project-Alias.

TOOLS
- » **3D printer (optional)** If you don't have one, you can access one at a makerspace or send the files out to a 3D printing service.
- » **Soldering iron**
- » **Wire stripper**
- » **Screwdriver**
- » **Computer with SD card reader**

TORE KNUDSEN is an interaction designer based in Copenhagen, Denmark, and works as a UX designer at Topp Innovation & Design in Malmö, Sweden. He holds a master's from K3, Malmö University.

BJØRN KARMANN is a senior designer at TellArt in Amsterdam, Netherlands. He holds a master's from Copenhagen Institute of Interaction Design and has won multiple awards for his graduation project, The Objectifier.

Bjørn Karmann and Tore Knudsen

WHAT IS ALIAS?

Alias is a teachable "parasite" that gives you more control over your smart assistant's customization and privacy. Through a simple app, you can train Alias to react to a self-chosen wake-word; once trained, Alias takes control over your home assistant by activating it for you whenever you say. When you're not using it, Alias makes sure the assistant is paralyzed and unable to listen to your conversations.

When placed on top of your home assistant (Figure Ⓐ), Alias uses two small speakers to interrupt the assistant's listening with a constant low noise that feeds directly into the microphone of the assistant. When Alias recognizes your user-created wake-word ("Hey Alias" or "Jarvis" or whatever), it stops the noise and quietly activates the assistant by speaking the original wake-word "Alexa" or "Hey Google" (Figure Ⓑ). From here the assistant can be used as normal. Your wake-word is detected by a small neural network program that runs locally on Alias, so the sounds of your home are not uploaded to anyone's cloud.

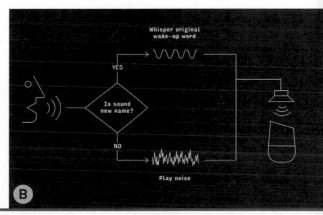

Ⓐ

Ⓑ

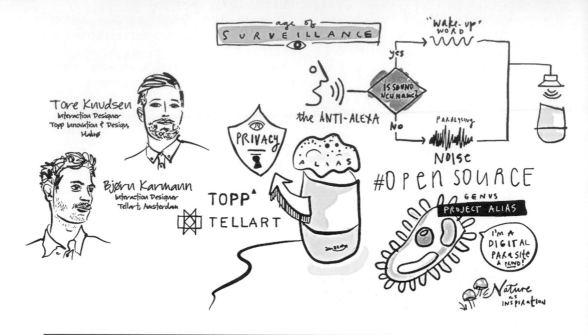

A YEAR OF ALIAS

When we launched Alias back in January 2019 we had no idea what an amazing journey we were about to step into. It was a hobby project we'd worked on during evening hours for 6 months — a fun challenge and a way to express some thoughts and values around our relation to tech, smart homes, and the future.

It didn't take long for online magazines and newspapers to catch on. Articles turned into interviews, and awards turned into exhibitions around the world. In the end, it was exactly what we hoped for: We never planned to bring another product into the world, but rather to inspire consumers to try an alternative solution and the idea that we, as the owners of our devices and our data, should have control over them.

Despite all the effort to make it a working prototype, Alias 1.0 was still not an out-of-the-box experience. The code was experimental and rough. So after popular feedback, we decided to rewrite the project from the ground up.

WHAT'S NEW (2.0)

When creating the early prototypes it became clear that there were no real-time sound classifiers fit for our purpose, so we decided to write our own

software for sound analysis: we turned the sound into spectrograms and used Keras (a popular TensorFlow wrapper) to make the machine learning logic. Our big ambition with this first approach was to allow any sound to be used, not just words. While it was fun to learn about all this, we never really got it to work better than a proof of concept. So now we're using an offline speech-to-text algorithm called PocketSphinx to do keyword detection (see "Open Source Voice for Makers," page 38). This has made Alias more reliable and easier to configure and train, relying on words from the dictionary.

Our solution of placing small speakers directly onto the microphone of the assistant inspired a lot of new ideas, and we realized that we could let the user configure custom shortcuts to tell Alias to send longer commands to the assistant. Previously we used an audio snippet of the original wake-word to wake up the assistant; our new design implements a text-to-speech algorithm (eSpeak) that makes it possible to not only change the wake-word but also specify exactly what commands to send. Now you can start a whole service from one wake-word. For example, you could say "Funky time!" and Alias can pass it along as "OK Google, play some funky music on Spotify." There's no limit

Karmann and Knudsen; illustration by Heather Lafleur

to how many wake-words/commands the Alias can be configured with at once (Figure C).

Having this new programmable voice that communicates with the assistant also made us think of different ways to create "data noise" to make the user more anonymous. So we've added a settings page where you can configure different parameters:

- We added a Gender option to let you choose what gender the assistant should perceive when Alias whispers command. By changing to the opposite gender you'll be able to introduce false labeling into the assistant's algorithm. This confusion might lead to interesting interactions and answers.
- We added a Language option to change the language of Alias when it speaks to your assistant — another layer of false labeling that makes the system label you with a different nationality. To use this feature, the command for the assistant has to be written in the selected language.

Besides that, we've also added options to optimize the performance of Alias, e.g. sensitivity, noise delay, and volume.

Follow the instructions at instructables.com/id/Project-Alias to build your new and improved Alias 2.0, and grab the latest code at github.com/bjoernkarmann/project_alias/tree/2.0. We've also provided a new *.img* file, so now installing the software on your Raspberry Pi is as easy as flashing an empty SD card.

A SMARTER FUTURE

Alias was one of the first public hacks to approach the privacy issues on home assistants, but since early 2018 many research and hobby projects have tackled the issue, such as the DolphinAttack method that sends voice commands by modulating ultrasonic frequencies you can't hear, or the recent Light Commands attack that uses lasers from a distance to manipulate the microphone.

It's clear this problem is something we care about but have yet to find the right solution. The future is hopefully going to be a decentralized one with private smart homes. But until that happens it's important to not accept the predefined lifestyle from Silicon Valley and to question whether it's really as "smart" as they say! ◉

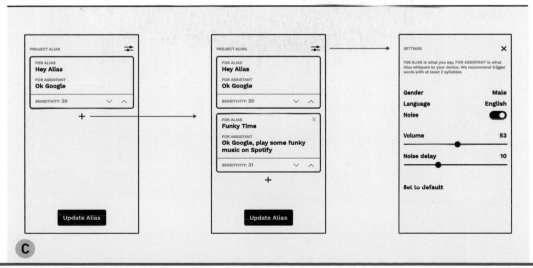

C

Your Own
Private
Smart
Home

Keep your IoT gadgets private and your data safe, with this DIY **Mozilla WebThings Gateway** on a Raspberry Pi

WRITTEN BY KATHY GIORI

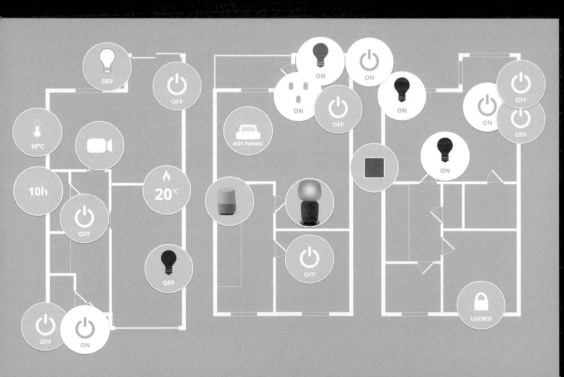

Smart home products, like dumb ones, fall into a couple different categories: there's "fun and entertaining" and then there's "practical." Fun stuff when made smart tends to offer me the benefit of convenience; for example, I can ask a smart speaker to play my choice of music. Practical products can bring me added safety and cost savings, such as fire and burglar alarms that notify me of trouble when I'm remote, detectors that sense a leak before it becomes a flood, or monitors on energy and water use, so I can tune my habits to protect our planet's natural resources.

You need money to make a change from dumb to smart, so the benefit of adding smarts to your home has to outweigh that cost. Another cost is the potential for unforeseen new problems. What downsides exist? The one risk universally discussed at Internet of Things conferences I attend is digital security. Unauthorized hackers exposing your IP camera feeds to the internet is super creepy. Imagine the dark web selling data on which homes contain latchkey kids who are alone after school, and which homes are empty during the workday. The potential losses and safety risks (foundation.mozilla.org/en/blog/securing-internet-things) are downright depressing.

Security is something that must be addressed by smart home products; that's a given. But in addition, I want to stress two additional important values that are often overlooked by today's smart home vendors: privacy and interoperability. Here's a short description of these three values, from my perspective:

» **PRIVACY** — For maximum privacy, I want my smart home system, and its data, located in my house. It's mine. I want to set it up. No company should be required to have access to it. It should be possible to add secure logins for everyone in my family. From within my house, the system should still work even if my broadband connection to the internet is down.

» **SECURITY** — A remote *https:* login is the best web security framework available. It's the same type of security mechanism I use to log into my bank account or pay for online purchases. I will have peace of mind knowing that my smart home keys only work using HTTPS, so that unauthorized hackers are locked out. Also, if none of my login keys or raw data are shared with a cloud data center, there's nothing to hack "up there" either.

» **INTEROPERABILITY** — I want to mix and match different smart home products and brands. And I don't want to be limited to searching for symbols of "works with brand A" (or B, C, or D). Instead, my brand-A pushbutton should be able to turn on and off my brand-B bulbs, my brand-C outlets, or anything else I've installed.

KATHY GIORI is senior product manager at Mozilla; in previous roles she promoted open source hardware and software at Arduino, Qualcomm, and various startups. She holds degrees in electrical engineering from the University of Minnesota and Stanford.

Mozilla Foundation

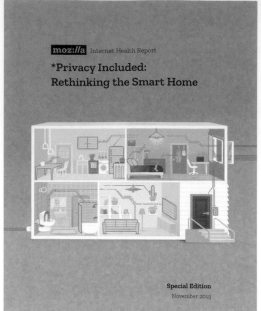

moz://a Internet Health Report

*Privacy Included:
Rethinking the Smart Home

Special Edition
November 2019

As outlined in the Mozilla Foundation's 2019 Internet Health Report (foundation.mozilla.org/en/blog/how-smart-homes-could-be-wiser), there's no single commercial vendor you can point to that does an outstanding job on privacy, security, and interoperability.

But with a maker spirit, and a mix of open source technology and off-the-shelf products, you can create a home environment that optimally addresses these core values. If you follow the approach outlined here, many benefits await:

- Your voice commands can be locally processed and therefore private.
- You can buy wireless devices with attractive designs and optimized battery life.
- You can easily make your own web things to mix with ones you buy.
- You can create custom rules for home automation tasks and remote notification alerts across a variety of brands.
- You can completely ignore smartphone apps specific to wireless Zigbee and Z-Wave devices.
- You can delete many of the brand-specific smartphone apps you've downloaded after getting a Wi-Fi product onto your local network (you only need them for first-time setup).
- You can "just say no" to cloud services accounts and ongoing fees.

MAKING YOUR DREAM HOME SMART — AND PRIVATE

"Where can I buy one?" you ask. Today, the solution described in this article is free and available to makers, early adopters, and anyone willing to roll up their sleeves to take advantage of personalized smart home control and monitoring. The maker approach is outlined in two parts:

Gateway —First you'll set up the central nervous system, the Mozilla WebThings Gateway. The gateway bridges all your smart device data to the web and lets you control and monitor everything from one personal web portal (Figure **A**).

There's a Getting Started Guide online and an excellent gateway setup video on YouTube that was created by Shawn Hymel.

Things — Then you'll connect your smart things to the gateway. It can be tricky to get good advice on what you need to buy and/or build, because your family and friends may have limited experience with smart home use. You typically have to search outside your comfort zone of local advisors. Fortunately, many smart home products can already be bridged to the WebThings gateway via its add-on framework (github.com/mozilla-iot/addon-list), due to impressive contributions from the open source community (Figure **B**). These include Apple HomeKit, Zigbee and Z-Wave, X10, Roku, Sonos, and many more; and for makers, boards like Arduino, Raspberry Pi, and ESP32 (for a list of supported hardware see github.com/mozilla-iot/wiki/wiki/Supported-Hardware). If you're a developer you can join the fun and tackle writing your own add-on to bridge an existing device.

Rather than having to reverse-engineer all sorts of different devices, it would be much better if the smart home industry could move toward a standards-based approach. The W3C Web of Things (w3.org/wot) is one such approach, and we continue to invest time in it at Mozilla. For lower-power devices that can't run a full web server on battery power, the recent "Project CHIP" announcement by Amazon, Apple, Google, and the Zigbee Alliance looks promising for improved interoperability of low-power IP-capable devices

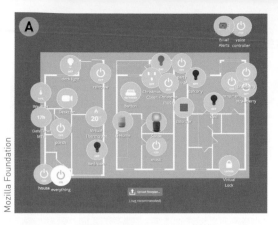

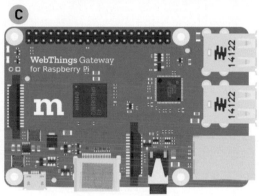

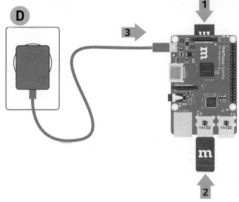

(connectedhomeip.com), It could tie into the Web of Things standard with a single add on, much as the Zigbee add-on does today.

We want to see standards that are open and accessible to all, like the internet itself. Then, using free and open source software and low-cost hardware, anyone could build their own native "web things" or buy from any vendor, rather than be locked into an ecosystem dominated by one or two brands.

BUILD YOUR OWN PRIVATE SMART HOME GATEWAY

1. CREATE A MOZILLA WEBTHINGS GATEWAY ON A RASPBERRY PI

The best approach we can recommend today is to build your own private gateway, and carefully select or build smart devices to manage through it.

It's easy to get started. To "self-install" the WebThings Gateway application onto a Raspberry Pi (Figure **C**), just download the Raspbian-based image from iot.mozilla.org/gateway, flash it onto a

microSD card, insert the card into the Pi (and any USB dongles If you want to talk directly to Z-Wave type devices), and boot the Pi (Figure **D**). That's it! You've built your own WebThings Gateway.

Another option is to use the Docker image (github.com/mozilla-iot/gateway-docker) or command-line installation (github.com/mozilla-iot/gateway) on any suitable computer (Linux, MacOS, or Microsoft Windows).

Complete the remaining steps in the Getting Started Guide at iot.mozilla.org/docs/gateway-getting-started-guide.html to set up Wi-Fi and add users to your gateway, and watch the setup video by Shawn Hymel at youtu.be/WHite8vdcbE. Then you'll likely need to install some add-ons for the devices you want to manage.

Follow the tips in the following sections to take advantage of a few of my favorite features. My favorite "convenience" add-on is the voice control add-on. It processes voice commands locally on the gateway. When I talk to my smart home, nobody is listening to me!

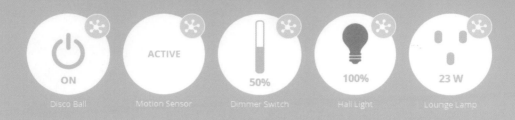

Disco Ball	Motion Sensor	Dimmer Switch	Hall Light	Lounge Lamp
ON	ACTIVE	50%	100%	23 W

2. CONNECT YOUR SMART THINGS TO MAKE THEM WEBTHINGS

Here's my starter list of useful smart home things, how I use them, and example products.

Get Connected

Smart Thing	Useful Traits	Product Examples
Light bulb	Convenient on/off, dimming, color coordination, and auto turn on/off based on whether a room is occupied	Ikea Trådfri, Samsung SmartThings, Philips Hue, Lifx, TP-Link, Cree, Sylvania
Power outlet (smart plug)	Fans, coffee makers, TVs, outdoor lighting strips, auto on/off to save energy	Ikea, SmartThings, TP-Link, Wemo
Door sensor	Normal coming and going, and intrusion detection (unexpected entry)	SmartThings, Xiaomi
Motion sensor	Occupancy for lighting automation, intrusion detection	Ikea, SmartThings, Xiaomi
USB "speaker-phone"	Microphone array integrated with speaker for private local voice commands, optional internet radio streaming	Jabra
ONVIF web camera	Checking on pets, kids, who's at front door, see what triggered a motion alert	Foscam
Door lock	Letting in guests, latchkey kids	Yale
Sonos speaker	Listening to music and podcasts	Sonos
Google Home smart assistant	Make your own text-to-speech (TTS) "announcements" or "intruder warnings" that can be triggered by rules (you can mute the device from listening, as I do). Eventually I think TTS should run on the gateway.	Google
Leak sensor	Under-sink leaks, water heater leaks	SmartThings
Thermostat	Control heating and cooling more efficiently	Centralite

3. SET UP USEFUL ADD-ONS

Pulse — I use one-second pulses to turn groups of things on and off. For example, in my own home I created a pulse called *house*, and another that is "inverted" called *everything*. I then create rules, one associating the *house* pulse "event on" with turning on a bunch of lights, and another associating the *everything* pulse "event off" with turning off those same lights. For added convenience without having to open a browser to the gateway nor pop open a smartphone app, I can simply speak a voice command such as "Turn on the house." Then all the lights in the *house* rule will go on at once. I can also say "Turn everything off" to turn them all back off.

Pulses are also great for grouping lights and assigning them all to the same color. I have a *red alert* pulse that triggers all the colored lights to turn red for 10 seconds. Right now I have it tied to the detection of an earthquake over 5.0 magnitude within 500km of my house, using data pulled into my gateway from the USGS via the Earthquake add-on (Figure **E**).

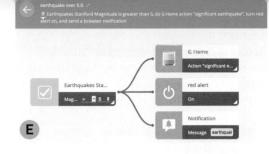

Voice control — Even though there's still a lot of room for enhancements and improvements to this add-on, including deeper integration with the gateway, it's already one of my favorites. I have several pulses that I use in conjunction with voice commands to turn devices on or off, and to trigger other actions. For example:
- "Turn on the rainbow" sets my smart bulbs to reflect the colors of a rainbow.
- "Turn on blueberry" sets them all to blue, and "strawberry" to pink.
- "Turn on the music" activates "playing" mode on my Sonos (Figure **F**), automatically playing whatever tune it was left pointing to as a source (i.e., streamed or local content).
- "Turn on the TV," "Turn on the coffee," and "Turn on the fan" are examples of devices I control using smart plugs.

There are no skills to install nor manual configuration required when adding new devices, adding rules, or changing the names of things. Your gateway figures it all out automagically. The commands just work, as long as the device you're commanding has a property you're allowed to

change, such as on/off. Slightly more advanced commands include "Set the study lamp to blue" or "Set the desk lamp to 50%" (for a dimmable smart bulb).

The same commands I can speak using the voice control add-on, I can also type into my gateway's web user interface. These typed commands are also processed locally and therefore completely private.

The current voice control add-on uses Snips libraries for the wake-word detection and speech-to-text elements of the voice stack (which can run locally on a Raspberry Pi 3). Since Snips was recently acquired by Sonos, and they announced that the open source components we use today won't continue to be innovated upon in the future, the voice team at Mozilla is researching how to replace those components with other open source options. We have already experimented and successfully integrated Mozilla's own DeepSpeech engine (for local speech-to-text processing) on the Raspberry Pi 4.

Zigbee — The sensors and actuators that I typically buy are Zigbee. I've found that this low-power radio protocol works well for door, motion, and leak detectors, as well as portable pushbuttons (all those things are battery-operated). A secondary benefit of Zigbee (and

Z-Wave) devices is privacy. They don't connect directly to the internet; they use non-IP protocols. By buying Zigbee light bulbs, power outlets, and other devices, I further enjoy the privacy of smart home devices that don't connect out to the internet without my permission or knowledge.

Notifiers — How can you hear an alarm or receive evidence of intruders when you're not home? Through a notification alert. Notifier add-ons are useful for letting me know about an important event or property change that I have configured in a rule to be monitored. I typically use browser alerts (since I'm at a computer a lot), but I have also tried email and messaging alerts. Many different notification add-ons that provide alerts are possible.

If I'm at my computer when earthquake data comes in, my Firefox browser will alert me and if I have the time, I'll go look at where it occurred and

how big it was (Figure **G**). The little ones occur multiple times per day.

Filtered internet content — One last type of useful add-ons don't represent data from smart devices installed in my home. Rather, I use them to pull data from the internet into my gateway (based on my configuration filters). I currently pull in date/time data (so I can trigger a front bedroom light automatically at sunset), earthquake data from the USGS (within 500km), local tide data from NOAA (since I row on San Francisco Bay), and local weather data (which I plan to eventually tie into my automatic sprinkler system).

Your thing data — whether locally produced or from internet queries — can be logged for up to 7 days. The logged data in Figure **H** show that many little earthquakes occur each day, within 500km of my house. (Three of them in the last day registered a magnitude over 3.0.)

Add-on curation — The first few add-ons (for Zigbee, Z-Wave, and native WebThings) were written by the Mozilla IoT team; since then, we've received far more add-ons from the open source community than we've contributed ourselves. Mozilla reviews and curates these add-ons and, once approved, makes them available to any user who wishes to extend the capability of their gateway. The full list of available add-ons can be found at github.com/mozilla-iot/addon-list/tree/master/addons or you can browse them (Figure **I**) by clicking the "+" icon on the add-ons page of a WebThings Gateway (they're listed alphabetically). Each add-on shows the author and the open source license. The author's name hyperlinks to the add-on's Github repository.

G

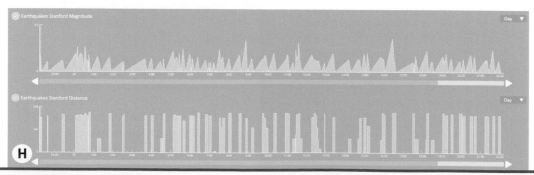

THE PRIVATE-BY-DESIGN APPROACH

Compare Mozilla's Smart Home approach to that typical of today's commercial solutions (Figure J). With WebThings, the data stay in the privacy of your own home, remote access is securely available via the internet using HTTPS, and all your smart products are made interoperable and manageable over a unified web portal.

HELP CREATE THE WEBTHINGS FUTURE

At Mozilla, we're still exploring the best way to build and distribute a smart home gateway with private local voice control and related products that support the needs of everyday users. We're eager for feedback and insight from the maker community, to help Mozilla bring "people first" IoT solutions to the next level. We hope you'll install the WebThings Gateway, optionally join our developer community, and let us know what you think! ⊘

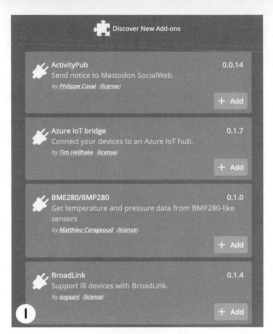

LEARN MORE:
- WebThings Gateway User Guide: iot.mozilla.org/docs/gateway-user-guide.html
- W3C Web of Things interoperability standard: w3.org/wot

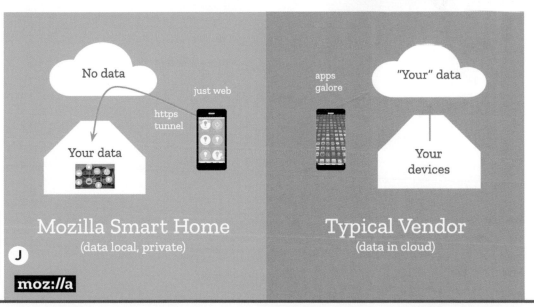

Mozilla Smart Home (data local, private)

Typical Vendor (data in cloud)

moz://a

Hey, Mycroft

Open source and private, this voice assistant works great on the Raspberry Pi

WRITTEN BY KRIS GESLING

TIME REQUIRED:
1–2 Hours

DIFFICULTY:
Easy

COST:
$60–$160

MATERIALS

» **Raspberry Pi 3 or 4 single-board computer** with appropriate power supply
» **microSD card** 8GB or larger
» **Microphone** The PlayStation Eye is a cheap and reliable option. If you're willing to spend more for a better experience, we recommend the Seeed ReSpeaker Mic Array v2.0, $64 from seeedstudio. com.
» **Headphones, external speakers, or dedicated amplifier** e.g. Logitech Z150 computer speakers or Adafruit's Class D MAX9744 amplifier board

TOOLS

» **Computer with SD card reader and internet connection**
» **Keyboard and monitor** for initial setup; not needed afterward

Mycroft is an open source voice assistant that respects your privacy. With Mycroft you can listen to music, check the weather, set timers or alarms, ask questions about the world, and much, much more. Unlike other assistants, however, Mycroft is the voice assistant you can trust.

Mycroft is private by default, and doesn't save any of your usage data unless you choose to share it anonymously to help the project to improve. Unlike other services, Mycroft is not using your data to sell you things, and because it's open source, you can see exactly what it's doing and be confident that it isn't listening when it shouldn't be.

You can buy a readymade Mycroft speaker or install Mycroft on a computer, but it also has a dedicated image for the Raspberry Pi, making it quick and easy to build your own device. You're free to remix, extend, or improve Mycroft however you wish. To this end, we've seen some very creative custom enclosures over the years, including integrations into full-fledged robots.

In this article we'll start with the basics, building your first voice assistant using just a Raspberry Pi and a microphone.

1. FLASH THE PICROFT IMAGE
Download the latest Picroft image from mycroft. ai/to/picroft-image. It's fairly large, 2.5GB, so you might want to grab a coffee while it downloads.

> **TIP:** For added security, you can verify the checksum of the downloaded image against the SHA256 hashes listed on our Picroft Github repository at github.com/MycroftAI/ enclosure-picroft#picroft---2019-07-20- stretch-lightning-release. If you don't already have a hash or checksum calculator, try the browser-based one at md5file.com/calculator.

Next, flash the image onto your microSD card. We recommend using Etcher (balena.io/etcher); it's free, simple to use, works every time, and runs on Linux, MacOS, and Windows. Once you've installed Etcher and plugged your microSD card into your computer, hit the Select Image button and point it to the Picroft *.zip* file you downloaded (Figure A). Double check that it has selected the correct device, then hit the Flash! button.

KRIS "GEZ" GESLING is director of developer relations for Mycroft.

A

2. CREATE A MYCROFT ACCOUNT
While you wait for your microSD card to flash, it's a good opportunity to create a Mycroft account at home.mycroft.ai.

You'll be asked if you'd like to contribute to Mycroft as a paid Member, or by opting in to the Open Dataset. Both are completely optional, and you can change your selection at any time. As a free and open source project, Memberships help cover the costs of running our servers and the wages of our small team who curate the software and ensure the privacy of your data. The Open Dataset is a free way to support our work by anonymously contributing your usage data that we can then use to improve our services.

3. SOFTWARE SETUP
Once Picroft is flashed to the microSD card, eject the card from the computer and insert it in the Raspberry Pi. On your first boot you'll see the bright red Raspberry Pi logo; the Picroft image is based on the official Raspbian operating system.

After a short period of automatic setup, you'll get to the guided Mycroft setup (Figure **B** on the following page). If your Pi isn't plugged into a network cable, Mycroft will guide you through connecting to Wi-Fi. Once connected, it will automatically search for any updates available for the Picroft setup and Mycroft's core software.

Next is the audio setup, starting with audio

```
MYCROFT
    Picroft

*************************************************************
** Picroft enclosure platform version: Buster Keaton - Pork Pi
**              mycroft-core: 19.8.5 ( master )
*************************************************************

Welcome to Picroft.  This image is designed to make getting started with
Mycroft quick and easy.  Would you like help setting up your system?
 Y)es, I'd like the guided setup.
 N)ope, just get me a command line and get out of my way!
Choice [Y/N]: []
```

```
============================================================
HARDWARE SETUP
How do you want Mycroft to output audio:
 1) Speakers via 3.5mm output (aka 'audio jack' or 'headphone jack')
 2) HDMI audio (e.g. a TV or monitor with built-in speakers)
 3) USB audio (e.g. a USB soundcard or USB mic/speaker combo)
 4) Google AIY Voice HAT and microphone board (Voice Kit v1)
 5) ReSpeaker Mic Array v2.0 (speaker plugged in to Mic board)
Choice [1-5]: 1 - Analog audio

Let's test and adjust the volume:
 1-9) Set volume level (1-quietest, 9=loudest)
 T)est
 R)eboot (needed if you just plugged in a USB speaker)
 D)one!
Level [1-9/T/D/R]: 9 - Saving

The final step is Microphone configuration:
As a voice assistant, Mycroft needs to access a microphone to operate.
Please ensure your microphone is connected and select from the following
list of microphones:
 1) PlayStation Eye (USB)
 2) Blue Snoball ICE (USB)
 3) Matrix Voice HAT.
 4) Other USB_microphone (unsupported -- good luck!)
Choice [1-4]: []
```

output (Figure **C**). In our case, we're using the 3.5mm audio jack on the Raspberry Pi, so we'll select option 1. Select the option that best fits your hardware. If you're using dedicated hardware like the Seeed Mic Array v2.0 or an AIY Kit, Mycroft will also update the firmware for those to ensure the best possible experience.

Now you set your audio output levels. Here you can play around with the levels 1 through 9. Each time you select a new level, Mycroft will change the volume and say "test." Unfortunately, the built-in audio output of the Raspberry Pi is quite underpowered, so you'll probably need to set this relatively high. Once you're happy with the volume, press D (for Done) to proceed.

The microphone setup provides a similar list of options as the previous output selection. If your specific microphone isn't listed, don't despair, it may still work, and we'll test the audio in the next step. If you chose the Seeed mic board for output, it'll automatically choose it for input as well.

To test your microphone and speakers, Mycroft will ask you to record 10 seconds of audio, then play it back (Figure **D**). If everything goes well, you can celebrate and continue! If you can't hear the test recording, run through the setup wizard again. Check the Audio Troubleshooting Guide in our documentation for the most common problems we've seen. And you can reach out for help in the Mycroft Community Forums.

Finally we have a few more advanced questions. For your first time trying out Mycroft, it's best to stick with the recommended settings using the stable and automatically updated "master" version of Mycroft (Figure **E**).

While writing this tutorial I'm on my secure home network, so I will keep the normal

Raspbian configuration of no password for **sudo**. I will however change the default password, as this can be used to remotely connect to the device.

Now the system will do some final setup based on your answers, and check for any underlying Raspbian system updates. Once complete, you can press any key to launch Mycroft!

4. MYCROFT CLI

Mycroft is a voice assistant, but sometimes it's really handy to see what's going on behind the scenes. This is why we have the CLI (Figure **F**) — a command line interface that provides a log output stream of what the system is doing, a list of the most recent spoken commands, an input line to type utterances or issue other commands to the system, and a mic level to visually see that Mycroft is receiving audio.

5. PAIRING THE DEVICE

In the CLI you can also see that your Picroft is attempting to communicate with Mycroft's servers. To do so it needs to be paired with your home.mycroft.ai account that you made earlier.

To ensure no one else can connect to your device, Mycroft will read out loud a six-character code. Everything Mycroft says is also output to the *History* panel. Keen eyes will also notice that the code is written in short form in the log output.

Add your device at home.mycroft.ai/devices/ addnow, using the pairing code provided by your Picroft, a name for the device, a description of the placement, and a geographic location so that it can provide the correct time, weather, and news station. Here you can also change the voice used by the device, and the wake word you'll use to issue commands (Figure **G**). You're all set!

```
Testing microphone...
When prompted, say something like 'testing 1 2 3 4 5 6 7 8 9 10'.
After 10 seconds, the sound heard through the microphone play back
for microphone verification.

Press any key to begin the test...
Already up to date.
Initializing...
Starting audiotest
========================= Info =========================
Input device: Default device @ Sample rate: 16000 Hz
Playback commandline: aplay -Dhw:0,0 WAV_FILE

========================================================
==        STARTING TO RECORD, MAKE SOME NOISE!        ==
========================================================
==        DONE RECORDING, PLAYING BACK...             ==
========================================================
Playing WAVE '/tmp/test.wav' : Signed 16 bit Little Endian, Rate 16000 Hz, Mono

Did you hear the yourself in the audio?
 1) Yes!
 2) No, let's repeat the test.
 3) No :(  Let's move on and I'll mess with the microphone later.
Choice [1-3]: []
```

D

```
 1) Yes!
 2) No, let's repeat the test.
 3) No :(  Let's move on and I'll mess with the microphone later.
Choice [1-3]: 1 - Yes, good to go
=================================================
MYCROFT SETUP
Mycroft is continuously updated.  For most users it is recommended that
you run on the 'master' branch -- which always holds stable builds -- and
allow the system to automatically upgrade with the biweekly releases.
 1) Use the recommendations ('master' / auto-update)
 2) I'm a core developer, put me on 'dev' and I'll manage updates
Choice [1-2]: 1 - Easy street, 'master' and automatically update
Already on 'master'
Your branch is up to date with 'origin/master'.
=================================================
SECURITY SETUP:
Let's examine a few security settings.

By default, Raspbian is configured to not require a password to perform
actions as root (e.g. 'sudo ...').  This allows any application on the
pi to have full access to the system.  This can make some development
tasks easy, but is less secure.  Would you like to remain with this default
setup or would you like to enable standard 'sudo' password behavior?
 1) Stick with normal Raspian configuration, no password for 'sudo'
 2) Require a password for 'sudo' actions.
Choice [1-2]: []
```

E

USE YOUR NEW PICROFT

Try a few voice commands to make sure Mycroft is working as expected.

Hey Mycroft, what time is it?

Hey Mycroft, how tall is the Eiffel Tower?

Hey Mycroft, play the news.

Hey Mycroft, why are fire engines red?

As you try more phrases, you might find that Mycroft can't complete your request. By default Mycroft can do a lot, but you can extend it by installing new Voice Skills from the Marketplace (market.mycroft.ai). Here you'll find a whole range of skills built by the Mycroft Community — skills to find cocktail recipes, connect to your smart home devices, play music, check your email, and many more. To install a skill, simply ask with your voice: *Hey Mycroft, install Cocktails.*

Chat to Movie Master to learn about different movies and get recommendations. Learn about specific Pokemon, their evolution, and their effectiveness against other Pokemon. Use the multi-sided Dice skill for your next tabletop adventure. Track the International Space Station and find out which country it's passing over.

If you have smart home devices, your Picroft can let you speak to them, rather than opening an app just to turn off a light. We have skills for systems including Home Assistant, OpenHAB, and the Mozilla WebThings Gateway (build yours on page 50). Just say, *Hey Mycroft, install Mozilla Gateway.* Then head to home.mycroft.ai/skills, select the skill, and click Connect.

WHAT NEXT?

Now you have your very own open source voice assistant. To learn about Mycroft technologies

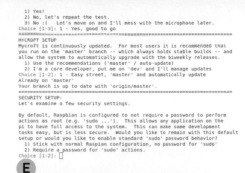
F

Configure your new device

Pairing Code *

Code spoken by device

Name *   Placement

Picroft   Kitchen

Geographical Location

Country   Region

Australia   Northern Territory

City   Time Zone

Darwin   Australia/Darwin

Voice

[British Male] [American Male] [Google Voice]

Wake Word

G [Hey Mycroft] [Christopher] [Hey Ezra] [Hey Jarvis]

and how to develop your own Skills, check out mycroft.ai/documentation. If you have questions or want to share, join the Mycroft Community Forums (community.mycroft.ai) or Chat (chat.mycroft.ai). Finally, we have so many ways you can get involved, including improving Mycroft's pronunciation, testing or writing new Skills, translating into new languages, or squashing bugs. Join the thousands of us at mycroft.ai/contribute who are contributing to open and accessible voice technologies for all humanity. ●

Face Jam

Keep clear from recognition algorithms with these methods

WRITTEN BY MIKE SENESE

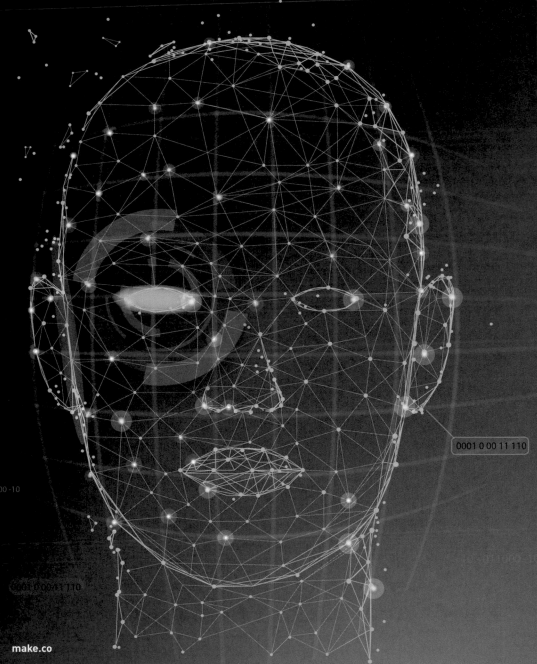

0001 0 00 11 110

011000 -10

0001 0 00 11 110

Hey, hey you... see those little black domes on the ceiling? On the light posts? They're everywhere. They're watching you. And they're smart enough to know exactly who you are.

Facial recognition is the reality nowadays, used broadly at events and concerts and even in the check-in line at CES. If you want to keep some semblance of anonymity in your life, we've put together a few options here that you can don. But be warned: Surveillance technology is secretive and algorithms are continuously tweaked, so there's no guarantee these will keep you perfectly private. And most will draw more attention to you than going without.

CAMERA CONFUSING CROWD PATCHES

EAVISE research students Simen Thys, Wiebe Van Ranst, and Toon Goedeme demonstrate that an image of a crowd held close to a person can counteract a person-detection algorithm, rendering someone "invisible" (arxiv.org/pdf/1904.08653.pdf) (Figure A). Shirts with their crowd images on them can now be found, as we saw at Maker Faire Shenzhen last year. You can get their code at gitlab.com/EAVISE/adversarial-yolo.

INSANE CLOWN POSSE MAKEUP

For those who have participated in the Gathering of the Juggalos, it turns out your homage to the face makeup worn by Violent J and Shaggy 2 Dope (Figure B) will help you confuse standard facial recognition programs, as deduced by twitter user Tahkion (twitter.com/tahkion/status/1013304616958607360). It won't, however, let you slip past three-dimensional systems like those on the newest iPhones.

BLINDING GLARES

With their IR-attuned reflective materials, Reflectacles (reflectacles.com) are glasses said to render cameras incapable of seeing their wearers' faces in day or night (Figure C). Probably the most discreet option, if they work as advertised.

OL' FASHIONED HOOD AND BANDANA

The most basic yet effective way to avoid facial recognition, especially if you conceal your eyes with sunglasses. But looking like a bank-robbing bandit sure can be suspicious (Figure D). ✐

Adobe Stock-lidiia, EAVISE, Insane Clown Posse, Reflectacles, Claudie Mcmichael

Boost That Bike!

Written and photographed by Justin Lemire-Elmore

TIME REQUIRED:
An Afternoon

DIFFICULTY:
Easy

COST:
$950–$1,000

MATERIALS

» **Bicycle** Just about any bike you have lying around, whether it's already a daily commuter or a vintage beater collecting dust in the garage. Old mountain bikes are usually ideal platforms, but almost any road bike, beach cruiser, or hybrid will work too.

One advantage of a front hub motor is that it doesn't matter what kind of gears or drive chain you have. The only important detail is that the front fork has ordinary slotted dropouts (not a modern high-end thru-axle system).

» **Hub motor conversion kit** Many suppliers offer conversion kits that normally include the motor, battery, and all electronic controls needed to run the system. We're using the G311 Front Minimal Ready-To-Roll kit from my company Grin Technologies, ebikes.ca/shop/ready-to-roll-kits.html. The kit includes a hub motor front wheel, battery pack, motor controller, torque arm, thumb lever throttle, and more (see "Your Basic E-Bike Conversion Kit" on the following page).

OPTIONAL:

» **Disc rotor** If your bike has disc brakes on the front wheel. The motors use a 6-bolt ISO disc that's readily available from bike shops.

» **Bottle-bob clamps** to secure the battery, if your bike frame doesn't have suitably located water bottle eyelets

» **Display** There are many options for handlebar displays if you want to view battery voltage, speed, and all that. But for this installation we're going to keep it minimalist.

TOOLS

» **Adjustable wrench aka Crescent wrench**
» **Allen keys, metric** All bicycles are metric!
» **Tire lever** or just a rounded flat head screwdriver
» **Wire cutters** aka side cutters, to snip spiral wrap and cable ties
» **File (optional)** may be necessary for motor axle to fit

JUSTIN LEMIRE-ELMORE got his first taste of e-bikes as an engineering physics student at the University of British Columbia and realized this was his life's calling. He's since devoted himself to the advocacy, science, exporation, and research of light electric vehicles. His company Grin Technologies makes and supplies conversion parts to the DIY e-bike community.

Convert any bike to electric with an easy front wheel motor kit — then zip around practically carbon free

Electric bicycles have been quietly taking the streets by storm in many parts of the world. Why? Because they're a fantastically fun and efficient way to move around. You get all the joys and freedom of riding a bike (avoiding traffic congestion and parking, for instance) while smashing through physical limits on your ability to climb hills and cover great distances. Once you've got an electric bicycle it's easy to leave the car parked in the driveway for all those short and medium-length errands. The e-bike will get you there faster, and with a 98% lower carbon footprint!

While there are many factory-made e-bikes these days, the movement was really pioneered and championed by a DIY maker community in the late 1990s and early 2000s who were electrifying their own bikes from scratch. You too can join in this phenomenon with just a spare afternoon.

In this project we'll show how to convert your own bike into an electric-assist bike, using a basic front wheel hub motor kit. There are countless options on the market for front motors, rear motors, and mid-drive motors spanning all kinds of weights and power levels, but we'll focus on a low-power front drive because it's among the simplest to install and has the best chance of being compatible with any random bike in your garage.

Extension cable

Cable ties

YOUR BASIC E-BIKE CONVERSION KIT

1 Hub motor wheel is small and lightweight (5lbs) and runs a totally silent helical gear system. It's not a power machine for racing up hills, but it provides enough boost to transform your bike into something new while still feeling and handling like a normal bicycle and not a scooter. This motor is already laced into a 700c rim, common on road and hybrid bikes. Other sizes are available, including 26" for older mountain bikes and 27.5" for newer ones.

2 Battery pack We chose a 36V, 16.5Ah downtube battery pack because it's light and still has decent range, about 35–40 miles on a charge. Larger batteries are available if you need to travel further, but they're proportionally heavier and more expensive.

3 Motor controller The Baserunner motor controller in this system is built into the

mounting cradle of the battery pack. On other kits the controller might be built into the motor, or supplied as a separate box.

4 Torque arm secures the motor axle to prevent it from spinning inside the frame. The motor itself has tabbed washers which do the job in many cases, but a proper torque arm provides extra security and is essential on weaker aluminum forks.

5 Throttle regulates the power you get from the motor. We're using a thumb lever throttle since it's most versatile, but there are also twist-grip throttles that act more like a motorcycle grip.

6 Spiral wrap and zip ties hold the motor extension cable and throttle wires snug against the frame and help the whole installation look neat.

INSTALL YOUR E-BIKE MOTOR

The following steps work the same for any ordinary bike. In this build, one of our new staff at Grin Technologies (hi Stuart!) had a used hybrid bike that he wanted to convert in order to speed up his commute.

You can watch a video of a similar kit installation at youtube.com/watch?v=t9jvgsOT6jo.

1. REMOVE THE FRONT WHEEL

The first step is to flip your bike upside down with it resting on the handlebars and remove the original front wheel (Figure A), which will be either quick release or threaded with nuts. If you have rim brakes on the bike you'll need to loosen the brake cable or deflate the tire to slide the wheel out from between the brake pads.

2. SWAP TIRE AND TUBE TO MOTOR WHEEL

Using the tire lever, or an improvised lever, carefully pry off the original tire from your front wheel being mindful not to mash the inner tube inside (Figure B).

Once the tire and tube are off, install them on the hub motor wheel by reversing that process (Figure C). Anyone who has fixed a flat tire will be familiar with this sequence, and it's usually our baseline question on whether someone has the comfort level to install their own kit.

3. INSERT HUB MOTOR IN FORK

Here's where things can get a bit challenging. The hub motor wheel should slide right into the fork dropouts if they have a 10mm opening, but sometimes the dropout slot isn't quite wide enough. If that's the case, use the hand file to enlarge the slot until the axle just fits.

Some kits use tabbed anti-rotation washers inside the dropouts to keep the axle from spinning; if these are used it's important that they're oriented with the tab pointing down. For this G311 kit, we're replacing the tabbed washer with a torque arm plate that provides much better spinout resistance (Figure D).

When you install the front motor, make sure the cable exit is rotated to point downward (Figure E) and the disc mount is on the left side of the fork (Figure F).

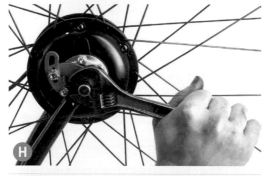

4. INSTALL THE TORQUE ARM

The torque arm is designed to screw into the fender eyelet hole that is present on the majority of front forks. It's held on with an M5 bolt, and a slotted channel allows it to adjust to different geometries and hole locations so that it lines up. For low-power motors like this, the torque arm isn't always required but it's a good safety measure to install one regardless, even with steel forks (Figure **G**).

5. TIGHTEN NUTS AND ADJUST BRAKES

Make sure the motor axle nuts are quite tight, since there's a lot of rotating torque on the axle (Figure **H**). (We recommend at least 40Nm or 30ft-lb if you're able to measure.) At this point, the motor is installed on the bike. You can inflate the tire and flip the bike right side up again so it's resting on the wheels.

Now make sure your front brakes are working again. If you have rim brakes, you might need to adjust the pad position to account for the width of the new rim. If you have disk brakes and installed a disk rotor on the hub, then adjust the caliper left or right so that it spins without rubbing.

6. MOUNT THE BATTERY CRADLE

The battery pack fits on the down tube of the bicycle, where it locks into a cradle that's held in place by the water bottle bolts. If your bike has mounting bosses (aka eyelets) for a water bottle cage that are in the right location (Figure **I**) then you're set, just use the supplied low-profile bolts so the head doesn't protrude (Figures **J** and **K**).

If your bike doesn't have water bottle eyelets or they're located too far back for the battery to fit, then you need to use the bottle-bob clamps. These attach securely to your frame tubing with hose clamps and allow you to position the battery anywhere you like.

7. INSTALL THE HANDLEBAR THROTTLE

In order to fit the thumb throttle on the handlebars you first need to remove the grips and sometimes the brake lever too. These grips have Allen screws for cinching to the handlebars; simply loosen the screws and slide them right off.

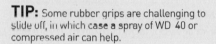

TIP: Some rubber grips are challenging to slide off, in which case a spray of WD 40 or compressed air can help.

The thumb throttle can be located on either the left or right side, and because it adds some extra width you may want to loosen and reposition the shifter and brake levers so that they can be reached comfortably without your hand feeling squished.

Slide the throttle into place and find a comfortable location for it and the brake lever and shifter (Figure L), then tighten them up and slide the grip back on (Figure M).

8. ROUTE THE CABLES

At this point, all the key hardware is on the bike and it's just a matter of plugging the parts together and securing the cable run. Route the throttle cable to follow one of the brake or shifter cables, using the spiral wrap tubing. It comes forward from the handlebar and then loops down along the down tube to the throttle plug on the motor controller (Figure N).

On the front hub motor there's an extension cable harness that links the motor to the motor controller. This cable runs up the side of the fork, held in place with cable ties, and then follows the down tube to the matching plug on the motor controller. Use cable ties as needed and snip off the offending pieces (Figure O).

9. PREFLIGHT CHECK

You're ready to slide the battery into the cradle (Figure P) and go for a rip! But before running the system, always double-check these details:

- The motor axle nuts are done up nice and tight
- You've reattached your front brakes
- The front tire has been inflated
- The battery key lock is locked ON so that the pack won't fall off when you hit a big bump.

With this done, you should be able to flip the battery's on/off button to the On position, see the controller LED glow steady red, press the throttle, and watch your life about to be transformed as the youthful joy of riding a bike returns in glory. Grin away!

TAKE YOUR FIRST E-RIDE

Your newly converted e-bike will easily cruise at 20mph on flat ground. It will go up modest hills (3%–6% grade) fairly well, but will struggle on really steep hill climbs unless you supply a decent amount of leg power to the equation. That said, it's still a *lot* easier than not having the motor.

The basic throttle control allows you to determine exactly how much power you want from the motor at any time. You can save the power just for hills and headwinds, or use it all the time to maintain a fast average speed. It's also possible to add torque sensors or pedal sensors to the bike frame so that the motor engages automatically when you turn the cranks, but those usually require more specialized bike tools to install.

RANGE FINDER

The range you can get from a given e-bike battery is no mystery, but it depends heavily on how it's used. Most people with a bike setup like this will consume about 15Wh/mile. That means a 500 watt-hour battery would have a range of 500 / 15 = 33 miles. Going up hills you'll need more like 30–40Wh/mi, while downhills consume nothing, and it averages out fairly consistent. If you don't pedal at all, your average consumption would be more like 20–25Wh/mi, while if you use the motor more sparingly you can easily get down to 8–10Wh/mi and have great range up to 50+ miles.

If you're purchasing a kit and want an estimate on how far it will go on a charge, it's best not to compare the claimed range, but just look at the size of the battery in watt-hours. ◉

LEARN MORE

Visit ebikes.ca and explore the Learn menu to find out why hub motors are awesome, the benefits of converting a bike over a factory e-bike, the history of the e-bike movement, and more. And jump into the e-bike community at endless-sphere.com/forums. Got a different kind of bike or different needs? Watch a video overview of all the different e-bike conversion options at youtube. com/watch?v=xnhuoUrwhIw.

Put More Watts to the Pavement

The future of hot-rodding is alive and bright in the EV world, from replacing a dirty two-stroke outboard with a quiet electric drive, to converting a full-size pickup with a tire-scorching 300hp electric. But where to start?

EV COMMUNITIES AND FORUMS

These are essential to get the latest information on components and techniques, and insight on your particular build. Connect with your local chapter of the Electric Auto Association or Electric Vehicle Association. Then check out www.evalbum.com, endless-sphere.com, visforvoltage.org, and diyelectriccar.com/forums for all kinds of EVs large and small, plus elmoto.net for motorcycles and electricseas.org for boats.

MOTORCYCLES

Building a kart or motorcycle is less intimidating than a car and much less expensive. But there's a big difference in the complexity and cost of an 80+mph freeway bike and an around-town motorcycle limited to 45mph.

There are now many different types of AC motors: *BLDC* ("brushless DC" — the controller runs on DC but the motor itself is AC), *PMAC* (permanent magnet AC), pure *AC induction* (no magnets), and the latest, *IPM* (internal permanent magnet). *PM brushed DC* motors are often overlooked as "old school" as they cannot "regen" well and require some maintenance, but they're often the lowest cost and best value in power for dollars, and they're by far the simplest to wire. They're excellent for first-timers.

Motorcycle kits from my company ThunderStruck Motors offer six different drives, from PM brushed (lowest cost) to BLDC and IPM drives. However, any one of our DC or PMAC motors (thunderstruck-ev.com/motors-dc-and-pmac) can be turned into a drive and, with the right controller, can offer mild to extreme performance.

CARS

For cars, the trend lately is to use salvaged components from production electric cars like the Nissan Leaf, Tesla, and others. Because it's often less expensive to buy a good used production EV than to convert a gas car, the trend to buy a junker and harvest all the batteries and the drive is really starting to take off.

Look to Stealth EV (stealthev.com) for their Tesla VCU (vehicle control unit), drives, and battery modules, and to Thunder Struck Motors for the Nissan Leaf VCU and for UQM drives. These VCUs allow the user to bypass most of the proprietary factory software that has prevented people from using these drives in other cars.

—Mark Schiess

MORE SMALL EVs YOU CAN MAKE

Mid-Drive E-Bike Conversion — makezine.com/2015/01/14/how-i-built-my-first-electric-bike

Power Wheels Race Car — makezine.com/2015/07/23/worlds-cutest-go-kart-2

Kick Scooter with Hoverboard Parts — instructables.com/id/DIY-Electric-Scooter-Conversion-With-Hoverboard-Pa

3D Printed Electric Unicycle — makezine.com/projects/3d-print-electric-unicycle

The Modular Fiddle

3D-print a bright-sounding, go-anywhere acoustic violin — and experiment with instrument design

Written and photographed by David Perry

TIME REQUIRED:
24 Hours (Print), 4 Hours (Build)

DIFFICULTY:
Intermediate

COST:
$80–$100

MATERIALS
- » **3D printed parts: neck, pegbox, body, bridge, and tailpiece** Download the free files for printing at openfabpdx.com/fiddle. I print these in carbon fiber PLA; its high stiffness-to-weight ratio creates a better sound. The printed tailpiece is optional; an off-the-shelf tailpiece may be used.
- » **Ukelele friction tuners (4)** The new pegbox design fits Grover 6-series tuners; I recommend 6B, Amazon #B00K8EAAIA. (An older version of the pegbox fits standard tapered pegs; it requires reaming with a pegbox reaming tool to fit properly.)
- » **4/4 strings, medium tension** D'Addario Ascente or similar
- » **4/4 violin tailpiece (optional)** Wittner #918119 or similar, if you're not using the 3D printed tailpiece
- » **Tailgut** Wittner nylon #Wit-1971 or similar, needed only for printed tailpiece. Nylon printer filament may also work for this!
- » **4/4 violin chin rest** Wittner #25011 or similar
- » **Carbon fiber tubes, 6mm×8mm×330mm (2)** Arris Hobby #AB0018, arrishobby.com

TOOLS
- » **Callpers or ruler**
- » **Small files** A triangular file is particularly helpful.
- » **Phillips screwdriver, small** to fit your tuners
- » **Deburring tool** These are much easier and safer than a knife for cleaning part edges.
- » **X-Acto knife**
- » **Needlenose pliers**
- » **Electric drill (optional)**
- » **Sandpaper** I use a 220 grit foam sanding block.
- » **Safety glasses**

DAVID PERRY is a mechanical engineer always seeking innovative ways to create things. In 2012 he attended the OSHWA Open Hardware Summit and was inspired to start a product design and manufacturing business, OpenFab PDX, using 3D printing, fun design tools, and open source practices. When he isn't driving CAD and 3D printers he's out mountain biking, gardening, and chasing a toddler in and around Portland, Oregon.

While there have been many 3D printed violins since my electric F-F-Fiddle in 2013 (see *Make:* Volume 40, makezine.com/projects/fffiddle), this one is truly disruptive. Previously, all 3DP acoustic violins either required assembly similar to that of a traditional instrument, or they had poor sound. The Modular Fiddle is absurdly easy to assemble and has good sound — significantly better than a cheap wooden instrument.

The Modular Fiddle is a 3D printed, open source, modular violin that can be produced with medium-size FFF 3D printers and off-the-shelf components (Figures Ⓐ, Ⓑ, and Ⓒ). It costs no more than $100 in materials and 4 hours or so in build time, and yields a good acoustic violin suitable for learning the instrument, solo playing, or playing in an acoustic jam.

Most important, the modularity provides a platform for experimentation and variation that's never been possible before. This is a test bed for innovation in string instrument acoustics.

The Modular Fiddle is absurdly easy to assemble and sounds better than a cheap wooden instrument.

THE VIOLIN, EVOLVED

Physical products are slowly entering a new era, in which the combined power of the internet and digital manufacturing allow widespread distribution and experimentation with product design.

The Modular Fiddle is designed to quickly test experimental violin concepts. Some acoustic innovations are only possible using 3D printing! Concepts that test well can be incorporated into traditional violin methods to advance the craft for all stringed instruments.

Violin history is rich with variations, and many elements of the instrument have evolved slowly over time. Luthiers attempted to copy other luthiers, but they sometimes made small mistakes that resulted in improved acoustics. Such mistakes led to the modern shape of the F holes.

What if we could accelerate that natural iterative process tenfold, a hundredfold, exponentially? We haven't seen a significant leap in violin design and construction since the invention of higher tension strings during the 18th century. We're due!

Using advanced digital design software and the Modular Fiddle, violin design and acoustics are poised to take a significant leap. 3D printing allows never-before-possible acoustic structure. Emerging design techniques, such as generative design (Figures **D** and **E**), could have significant impacts on instrument weight and performance.

Despite all this, the wood used for violins has incredible properties, like its stiffness-to-weight ratio. The violin is a finely tuned, highly evolved, powerhouse machine for producing beautiful, clear, and projecting tones. Can a 3D-printed fiddle possibly match this quality? No — and it shouldn't try to do so.

GOOD SOUND IS SUBJECTIVE

What we consider to be "good sound" changes over time. Desirable instrument sound is a combination of musical style, human psychology, and culture.

Currently, the Modular Fiddle has the sound quality and playability of a mid-range wooden violin with a noticeable decrease in volume and projection. While the current sound quality is

A lightweight, futuristic pegbox and bridge created using Autodesk's new Generative Design tools in Fusion 360.

Curious about what it's like to play a Hardanger? Now you can pop a new neck on your Modular Fiddle to find out!

A 5-string version of the Modular Fiddle is also available.

impressive, our focus should not be on matching the quality of wood — instead we should try to create altogether different sounds.

MAKE OLD FIDDLES NEW AGAIN

The modularity of this fiddle encourages experimentation. We can remix old instruments, like the Norwegian Hardanger fiddle or hardingfele. It's got a long and colorful history and a very different sound. The Hardanger Fiddle Association of America says:

> Its most distinguishing feature is the four or five sympathetic strings that run underneath the fingerboard and add echoing overtones to the sound. The traditional playing style is heavily polyphonic. A melody voice is accompanied by a moving "drone" voice. Together, the instrument and the playing style create the sound for which the Hardanger fiddle is famous.

The first major variation designed for the Modular Fiddle is a Modular Hardanger Fiddle (Figures **F**, **G**, and **H**). This instrument uses Hardanger strings and understrings to provide a Hardanger sound and playing experience.

BUILDING YOUR MODULAR FIDDLE

To build your own Modular Fiddle, you can download the parts lists and the design files for printing, and find complete build instructions at openfabpdx.com/fiddle. The 4-string version

shown here is completely free and open source, while the 5-string (Figures **I** and **J**) and Hardanger files are available for $12.50. With the download you get both STL files and STEP files, allowing easy remixes no matter your CAD preference.

If you'd like to buy a kit of parts, or a fully built Modular Fiddle, check out openfabpdx.com/shop.

There's also an active Google Group that supports all of OpenFab's violin designs. Makers of both the Modular Fiddle and its predecessor the F-F-Fiddle should head to the OpenFab Violin Builders group to review discussion and ask questions.

WHAT'S NEXT

I'm currently working to develop a new version of the 3D printed Modular Fiddle that's resistant to creep and to the effects of heat. Initial tests have shown huge improvements by using annealed Proto-Pasta HTPLA-CF filament, a high-temperature, carbon fiber-reinforced PLA that can be heat treated after printing for extra strength. You can read more about my progress at blog.proto-pasta.com/anneal and blog.proto-pasta.com/hardanger. ◑

Lathe Lessons Learned

**Written and photographed by
Stan Adermann and Vern Adermann**

How we converted a Jet metal lathe to CNC using free LinuxCNC software — and a few hardware misfires

TIME REQUIRED:
Many Weekends

DIFFICULTY:
Advanced

COST:
$2,500–$2,800

MATERIALS
» **Jet metalworking lathe with digital readout (DRO)**
» **3-phase motor** Marathon #145THFR5329, $500, to replace a burnt spindle motor
» **Variable Frequency Drive (VFD) motor controller** Emerson Commander SK, $450
» **CNC control board, parallel port** such as C11G Multifunction CNC Board, $68, CNC4PC.com
» **Stepper motors, NEMA 34 (2)** for X- and Z-axis, Model 34HS38-3008S, $110 each
» **Digital stepper driver boards (2)** GeckoDrive G213V, $150 each
» **PC running Linux CNC software** free download from linuxcnc.org. We used an ancient Pentium 4 machine.
» **Roxburgh EMC filter** for noise reduction
» **Ball screw, 40", with ball nut** $225
» **Thrust bearings (4)**
» **Custom motor mounts (2)** made from stainless and aluminum on our Tormach 770 CNC mill
» **Beam couplings (2)** aka flexible shaft couplers, $5 to $50 on Amazon depending on size
» **Control box, steel, 24"×16"×10"**
» **Switches** for power, safety disconnect, etc.
» **Wire: 12ga, 14ga, and 22ga**
» **Salvaged relay, switches, etc.** from parts of the lathe not used anymore

TOOLS
» **CNC mill, end mills, boring bars, turning tools** to machine the motor mounts
» **Drill, screwdrivers, wrenches, wire strippers, crimpers, etc.**
» **Soldering iron and solder**

STAN ADERMANN is a lead firmware engineer in Xbox hardware with 20-plus years' experience at Microsoft.

VERN ADERMANN is CEO/founder of CityWide Co. and an engineer with 20-plus years' experience in robotics and process systems in the semiconductor and aerospace industries.

Our home shop includes a Jet GBH-1340A metalworking lathe with a Digital Read Out (DRO), and we'd been discussing adding CNC control, because certain types of parts are extremely difficult to make with high precision without a computer in the driver's seat.

But we approached the project with a certain amount of procrastination. We initially selected a Variable Frequency Drive (VFD) controller for the spindle, NEMA 34 stepper motors, and drivers for the lathe axes based on what we found inside our Tormach 770 mill. We also found a parallel CNC control card online. One of our primary considerations for the parts we selected was "cheap," although in the end that would cost us.

The parts arrived and we set them aside for about a year while other projects took priority, only infrequently revisiting this project to take measurements or ponder how exactly to mount the steppers. It was the abrupt failure of the Jet's spindle motor that brought this project back into focus. We pulled out the parts and started the conversion in earnest.

There are three main parts to a CNC conversion. They are: (1) modification of the machine itself, (2) construction of a control box, and (3) setup and configuration of a control PC.

Parallel control board/interface card.

A Our 40" metal lathe, pre-conversion.

LATHE MODIFICATION, PART ONE

Our lathe has a 13" swing and a 40" long axis (Figure **A**). Spindle speed is normally controlled via a gearbox behind the spindle, driven by a single-phase 230V motor. No modification to the gearing was required; we simply selected an optimal gear setting and under the CNC the speed would be controlled by the VFD controller. The failure of the original single-phase motor was actually a stroke of luck, because replacing it with a three-phase motor provided more control and allowed for a top speed nearly double the 1,750 RPM the original motor could manage. Best of all, the VFD was able to convert 220V from single- to three-phase. The original electrical control box was removed from the back of the lathe and some of the control relays and other parts cannibalized for use in the new control box.

The carriage that holds the cutting tools had two options for controlling its Z-axis movement. (On a lathe, the Z-axis is left-right while the X-axis controls the diameter.) There's a primary leadscrew for general-purpose cutting, and a second leadscrew that rotates in lock-step with the spindle for cutting threads. Both of these are driven from the same gearbox, and engaged to move the carriage via control levers on the carriage itself. We opted to remove the thread-cutting leadscrew and the bar that controlled the primary leadscrew. This would allow us to drive the primary leadscrew via a stepper motor mounted at the opposite end, and attached by a belt and pulleys. The primary leadscrew took

just over 50 rotations to move the carriage 1" and we hoped this would provide some degree of precision control. Using a CNC mill, we crafted a motor mount to bolt to the lathe on a swivel, much the way the alternator in a car is mounted to allow belt tensioning (Figures **B** and **C**).

For the X-axis, aka the cross slide, the obvious choice was direct drive by the stepper. We removed the hand cranks from the machine and milled another aluminum mount. The leadscrew was attached to the stepper via a beam coupling to alleviate stress (Figures **D** and **E**).

We made no modifications to the lathe tailstock. This will remain under manual control while the computer handles the heavy lifting for the X and Z axes.

CONTROL BOX CONSTRUCTION

The original Jet control box was far too small to fit all the components required to control the lathe, so we found a 24"×16"×10" box online that would fit everything. A 10" depth was perhaps overkill, but gave ample room for mounting cooling fans and switches on the side of the box. And it was sturdy enough that when mounted on back of the lathe it could support a monitor and stand.

We mounted the components on a ⅛" sheet of aluminum which could be removed from the box for easy access, and which also would act as a heat sink (Figure **F**). Holes were cut in the aluminum and the back of the box for a set of manual spindle controls. Open slot wiring raceways were added to keep the wires from

First attempt at Z-axis motor drive.

First Z-axis motor mount being milled.

Replacing the cross slide: Original X-axis hand crank.

Cross slide motor assembled: the new X-axis stepper.

Laying out control parts.

Carefully labeled junction.

becoming a tangled mess (Figure Ⓖ).

Throughout the process a schematic was maintained in Visio with all connections carefully numbered, and the wires were labeled at both ends to match (Figure Ⓗ). Altogether, assembly of the control box took about 60 hours (Figures Ⓘ and Ⓙ).

Wiring in progress.

Assembled control box, first testing.

Assembled control box (with Stan's head inside).

SETTING UP A CONTROL PC

While many CNC projects use a parallel port to drive the machine, they often don't use the latest and greatest PC hardware. First, many modern PCs don't have parallel ports, but also many modern processors have optimizations which make them very good at running software but very bad at directly bit-banging I/O ports for time-sensitive hardware control. It's not a problem for a PC driving a printer because USB has offloaded the hard work, but in our experience with a CNC mill, the wrong hardware/software configuration can result in a cut being made tens of thousandths off from what the G-code asked for.

Fortunately, the major options for CNC software have lists of computers that have been tested, so less guesswork is required. We opted for an old Dell Optiplex with a Pentium 4 processor running LinuxCNC. We were able to purchase two of these (so we had a spare) for $30 apiece at a local recycled PC store.

LinuxCNC (linuxcnc.org) offers a very capable set of control options and is well supported by a community of dedicated computer geeks. Following the instructions from the website, installing LinuxCNC was straightforward and it ran well on our ancient PC. Using their StepConf program we were able to configure individual pins of the parallel port any way we wanted. We discovered it would have been worthwhile to set up LinuxCNC before we had purchased any of our control hardware. They had default configurations for several different types of hardware, some of which we'd been unaware of when we made our original purchases.

It didn't take long and our parallel control board was lighting up like a Christmas tree when we clicked buttons. No magic smoke came out. We should have been golden. Instead, nothing worked.

CHEAP PARTS COME BACK TO BITE US

It's perhaps not fair to say nothing worked. There were hints that some things were almost right. One of the stepper motors would make a single *chug* noise when we told it to turn. The LED on one of the stepper drivers would stay green up until that point, and then it would turn red. The other stepper driver defiantly went red as soon as power was applied and sat there glaring at us like the eye of Sauron.

We reviewed all our wiring. We compared our setup to how the Tormach was wired; no problem there. It was when we borrowed an oscilloscope to check the output of our CNC control board that we found the first problem: the output signal voltage was only rising to half the level required by the stepper drivers. Our $20 control board was garbage. We decided to loosen the wallet a bit and found a $99 board on another website. When it arrived, it was branded with a different website: CNC4PC.com. It was also six hardware revisions out of date from their newest offering. It did provide the right voltage levels, so we called it "lesson learned" and hoped the steppers might work better. They did not.

I mentioned earlier that much of what we put in our own control box was based on what we found in our mill. These stepper drivers were the same model MA860H as the mill had, so with visions of mill repair bills in our heads we started swapping our suspect parts into the mill. The stepper motors went first and to our great relief they both worked perfectly. The stepper drivers went next, and neither would function. The eye of Sauron continued to taunt us. Suspecting this might have been our fault we ordered a new pair of the same model. Both of these were dead on arrival as well. One would not function in the mill at all and the other would turn but only in one direction. Clearly these drivers were not a reliable solution.

More online research brought us at last to GeckoDrive, and we could not be happier. These boards worked well the first time we hooked them up, and both of the Gecko units together fit in the same space as one of the earlier, cheaper drivers. At the voltage required, the GeckoDrives would need a heat sink (not included). The cheaper drivers included a large heat sink and fan, and since the fan was the one part of that driver that worked reliably, we found much satisfaction in mounting the GeckoDrives in the hollowed-out carcass of the dead driver (Figure **K**). This solved two problems at once. The new parallel control board was much larger than the original and now we could mount it in the space left by one of the original stepper drivers (Figure **L**).

One note about stepper drivers: They come in

K

Frankenstein motor driver: New GeckoDrives mounted in carcass of failed motor driver.

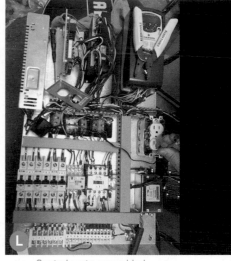

L

Control parts assembled, not in housing yet.

analog and digital flavors. On YouTube you'll find videos comparing their behavior. These show, and we agree, that digital drivers provide much smoother and quieter stepper motor operation. They seem well worth the added price.

PUTTING IT ALL TOGETHER

All the pieces were in place. We could control the steppers with UI buttons or G-code instructions, and with a rudimentary attachment of the motors to the leadscrews we could move the carriage along both axes.

We didn't know the exact ratio of leadscrew rotations to lateral movement, so it took a bit of trial and error to find the right settings in the LinuxCNC StepConf program. StepConf asks you for several values: motor steps per revolution, driver microstepping, pulley teeth ratios, and leadscrew pitch. If you're uncertain on these values, it's worth noting they just get multiplied into a single value that means "steps per inch." If you set all but one of these values to 1 (doesn't matter which), you end up with the remaining value being a large number that you can tweak with great precision. The process we followed:

1. Move the carriage to an approximate known position, moving from left to right. In your CNC UI, reset your offsets so the position reads 0.
2. Measure the position of the carriage.
3. Using G-code, move the carriage 1" farther to the right, i.e. **Z1**.

4. Measure the new position of the carriage and calculate the difference in inches.
5. Divide your "steps per inch" value by the distance moved, for a new steps per inch value. For example, if your steps per inch is **20000** and you move 1.015", your new steps per inch would be 20,000/1.015 or **19704**.
6. Repeat, until issuing a command to move 1" reliably results in 1" of movement.

It's critical that your measurements always be made after moving the carriage in the same direction because your leadscrew will almost certainly have some backlash. If you measure after moving in opposite directions, your measurements will be off by up to the backlash amount.

The DRO was still attached to the lathe and this greatly simplified the process of comparing instructions on the PC to the actual carriage

M

Measuring where to mount a Z-axis motor mount.

Our second Z-axis motor mount: midway through milling, completed, and mounted with ball screw.

movement. Following our process, we should have arrived at a steps per inch value that would give consistent results no matter where on the axis we measured. It worked for the X-axis, but along the Z-axis measurements varied by as much as 0.012" depending on where they were taken. Something was horribly wrong.

LATHE MODIFICATION, PART TWO

Leadscrews can be inaccurate, but the screw would have to be very bad to have the error increase and decrease again and again over 40". The problem was that in addition to the headscrew, there were other gears and worm drives involved in Z-axis movement. We needed to account for inaccuracy across the entire train. The Z-axis backlash was similarly awful. LinuxCNC has ways to compensate for this, but it would have required finding the error at every point along the 40" axis. It would have been nearly impossible to achieve the precision we wanted. The gear train would have to go.

A precision ball screw can virtually eliminate backlash, for a price. One company quoted $3,500 for a 40" ball screw. In the end, we purchased a ball screw and nut for $225 from Roton Products in Missouri. It required modification to fit the bearings we'd purchased, $336 at a local grind shop. The Roton ball screw had 0.007" backlash, but at least it was consistent and easily compensated for in LinuxCNC.

We also decided to eliminate the belt and pulleys and create a new mount for the ball screw (Figures M, N and O) so that we could do direct drive from the stepper motor. Each end of the

Z-axis ball screw headstock mount.

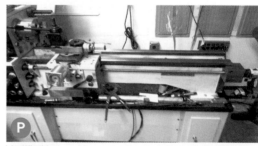

Modified lathe with new motor mounts ready for steppers.

Mounting a limit switch.

shaft is held by a pair of thrust bearings mounted back to back to eliminate movement while still permitting rotation, and the shaft itself is mounted with a bit of tension between the two bearing mounts (Figure P).

Any CNC needs limit switches so the machine can find home position for each axis. Fortunately, when we dismantled the manual control box we found two momentary switches perfect for this purpose, so we mounted them in a convenient position for each drive screw (Figure Q).

For cabling you would normally have cable tracks, but we avoided this for the X-axis by simply allowing the wires going from the control box on the back to the X-axis driven from the front to hang loosely under the machine.

FINISHING TOUCHES

We now had a functional CNC lathe (Figure R). LinuxCNC was working well, although the UI looked like an old Windows 98 app. Fortunately, we found two alternate user interfaces after perusing their user forums, both of which looked and worked much better (Figure S).

As is usual with Linux, be prepared to read lots of forums and documentation and to edit text files to get the configuration you need.

We also plan the following future enhancements:

» We lost the ability to cut threads with this project. However, LinuxCNC supports this if you can provide feedback from an optical spindle speed sensor.

» Liquid coolant can be very useful, even on an uncovered lathe at low RPMs.

» We can limit backlash by ordering new ball nuts where every fourth or fifth ball is a different size to tighten the tolerance between the ball screw and nut.

» The ball screws need to be protected. We need to craft some covers or brushes to keep them clean. ◐

The CNC lathe running, in its first test run.

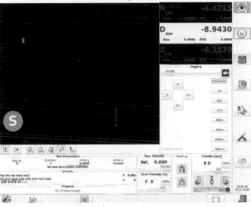

LinuxCNC screenshot (no program loaded until I figure out how to make it ignore the fact it's not hooked up)

A jig for shaping and sharpening knife blades. Handles made on the CNC lathe!

1+2+3 An Adorable Fuzzball

Written and photographed by Amanda Formaro

Help save the galaxy with this easy, pocket-sized Wookiee

TIME REQUIRED:
45 Minutes

DIFFICULTY:
Easy

COST:
$10

MATERIALS
» Stirring stick for gallon-sized paint can
» Faux fur, brown
» Felt, dark brown
» Felt, gray
» Pom-pom, medium, brown
» Googly eyes (2)

TOOLS
» Hot glue gun
» Scissors

AMANDA FORMARO
is the crafty, entrepreneurial mom of four and author of *Star Wars Mania*, a fun craft book for kids. She also runs a popular craft blog at CraftsbyAmanda.com

FUN STAR WARS FACTS:
- Chewbacca is a Wookiee; yes, there are two e's and the species name is capitalized!
- Did you know that George Lucas' inspiration for the Wookiee was actually his dog, Indiana? George Lucas has said, "*He was the prototype for the Wookiee. He always sat beside me in the car. He was big, a big bear of a dog.*" (And yes, Indiana the dog also influenced the naming of Indiana Jones.)
- Wookiees come from the planet Kashyyyk, average almost 7 feet tall, are comfortable in most climates, are excellent marksmen, and live for several hundred years!
- While Wookiees speak a language of their own (called Shyriiwook), they are capable of understanding humans. They are unable to speak our language due to the structure of their vocal cords.

If you're out looking for a Chewbacca craft, you won't find a ton, at least not yet. But as far as characters from *Star Wars* go, Chewbacca and R2-D2 seem to be the fan favorites. When I was coming up with the list of ideas for my *Star Wars* craft book, Chewbacca was on there several times. My favorite Wookiee craft from the book was this paint stick Chewbacca; I just think he turned out so cute!

1. Cover the paint stick with faux fur. It's easiest to turn the fur upside down and glue the stick to the back (Figure **Ⓐ**), then glue, fold and trim. Then cut two 5"×½" strips from the remaining fur for arms (Figure **Ⓑ**). Glue arms to the back of the stick.

2. Cut a 9"×½" strip of dark brown felt for the bandolier, and six squares and rectangles from gray felt for the ammo boxes (Figure **Ⓒ**). Glue the ends of the brown felt together to connect them and glue the ammo boxes onto the bandolier (Figure **Ⓓ**). Place bandolier over the top of the paint stick, over one arm and under the other. Glue it in place.

3. Glue on the pom pom nose and googly eyes.

I'm not going to lie, this was a super messy craft because of the faux fur. It's fine when you buy it, but as soon as you start cutting it it's just like being at the hair salon, only without the plastic cover they put on you. Little hairs. Everywhere. So if you want to adapt your Chewbacca craft and use felt instead, you won't get that furry feel, but he'll still be cute and you won't have to vacuum. Probably. ✪

The Negotiator

Written and photographed by Tyler Capps

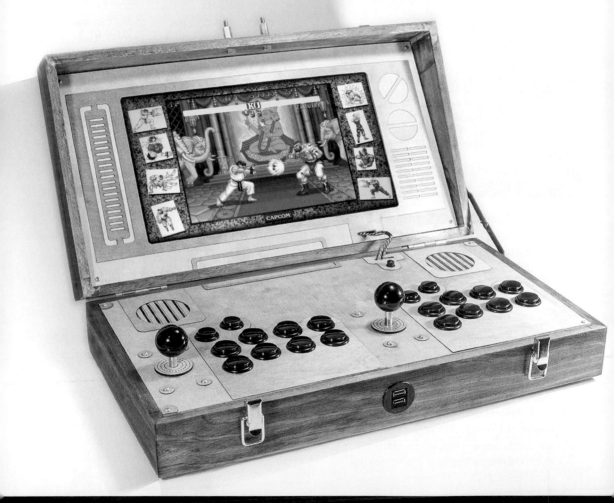

TYLER CAPPS is a freelance illustrator and designer (cappslock.com) in Asheville, North Carolina. He recently shifted his focus to woodworking and electronics to pursue building unique, finely crafted retro gaming devices at endbosscustoms.com.

TIME REQUIRED: 40 Hours

DIFFICULTY: Moderate

COST: $400–$600

MATERIALS
- » Raspberry Pi 3B+ single board computer
- » microSD card, 64GB
- » Power cable, Micro-USB
- » USB extenders, 6" (2)
- » LCD screen, 15.6"
- » LCD controller board with 12V 1A power supply
- » Power supply, 12V 1A for LCD controller
- » HDMI cable, short
- » Stereo speakers, desktop PC type
- » Speaker grill cloth
- » Extension cord, 3 plug
- » Power switch, arcade cabinet
- » Wire disconnects, 14–16 gauge, female (5)
- » Power cord, IEC C13
- » Arcade controller parts: joysticks, buttons, controller boards, and wiring
- » Boards, pine
- » Plywood, birch
- » Plywood sheets, birch, ultra-thin
- » Pre-stain conditioner
- » Red oak stain
- » Spray lacquer
- » Spray adhesive
- » Hinges (2)
- » Latches (2)
- » Lid support hinge
- » Rubber feet (4–6)
- » Various screws (lots)

TOOLS
- » Screwdriver
- » Hand drill
- » Soldering iron
- » Wire strippers
- » Heat-shrink tubing
- » Heat gun
- » Electrical tape
- » Hot glue gun
- » 3D printer
- » Computer with CAD software I used Tinkercad.
- » Thickness planer
- » Saws: table saw, band saw, and handsaw
- » Drill press with Forstner and other miscellaneous bits
- » Sanders: belt and random orbital
- » Sandpapers, 80–220 grit
- » Laser cutter
- » Orange Clean cleaner/degreaser
- » Paint brushes, foam
- » Rags

Carry this fully functional arcade machine anywhere in its integrated, foldable case

Not long ago my brother asked me to help him convert a glass coffee table into a cocktail arcade machine. I was already familiar with RetroArch and emulation in general, but this was the first time I worked with a Raspberry Pi. It was so easy to set up, and so much smaller than I expected, that I thought to myself, "I could do more with this." That build was pretty simple, but it gave me an itch I could not resist. I resolved to build a system into a disused suitcase that could be folded up and carried.

I kept an eye out for an appropriate folding case, but after a couple of weeks of looking I was at a dead end. Every box I came across was too small, too big, too shabby, too expensive, too thick, too thin, and so on. I realized I would probably just have to make it myself. Thinking more about it and drawing out a design in my head, I found myself with a list of design requirements:
- Foldable
- Light enough to carry
- Small enough to move easily
- Big enough to be functional
- Well-made and well-finished wood construction

This was becoming a pretty ambitious project given that I had no tools of my own, no experience working with wood, and only a passing knowledge of electronics. Fortunately there happened to be an amazing makerspace really close to me called Reforge Charleston (reforge.io). After joining, and with guidance from a few of the members (and a lot of YouTube videos), I taught myself basic woodworking, how to use the 3D printers, laser cutters, CNC machines, and everything else I would need to make my arcade box. Here are the main elements of what it entailed.

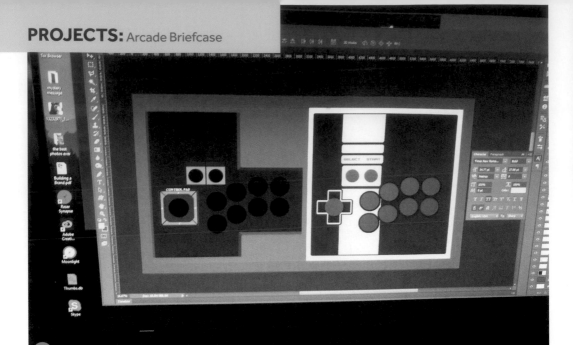

A

1. I used the measurements of the arcade controller parts I had selected to design the layout of my buttons and joysticks in Adobe Photoshop and Illustrator (Figures **A** and **B**). The layout out of these controls, accounting for speakers and space for the other electronics, gave me the dimensions of the box I needed to make — 21"×11¾"×5⅜".

2. I made the box using pine boards from Lowe's (Figure **C**). I constructed the frame first, then ran it through a table saw to cut it into two pieces to make the base and lid of my box.

3. I glued in the top and bottom boards; once dry, I sanded it all from 80 grit to 220 grit (Figure **D**).

4. I used a pre-stain conditioner, red oak stain, and many layers of spray lacquer to finish the box (Figure **E**). I know many woodworkers frown on using stain, but I opted for it because I wanted to keep costs down and also I didn't yet trust myself not to ruin perfectly good walnut.

5. Using the layout designs I had created in Illustrator, I laser-cut the control face from birch

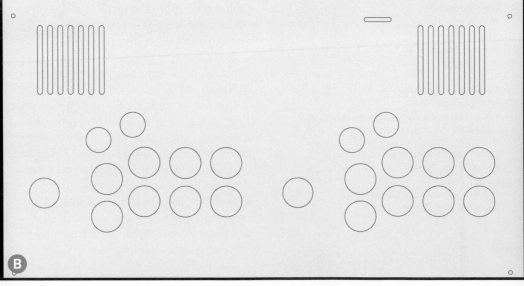

B

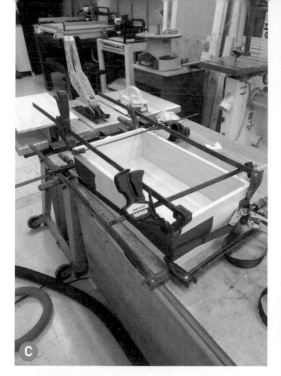

D

E

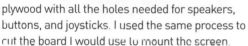

C

plywood with all the holes needed for speakers, buttons, and joysticks. I used the same process to cut the board I would use to mount the screen.

I went back to Illustrator and designed the overlays for the control and screen boards, as well as the *Pac-Man*-inspired dust covers for the joysticks. I laser-cut all those overlay pieces from ultra-thin birch sheets, then I used spray adhesive to secure them over the plywood faces. All the plywood was pre-sanded, so I coated them in many layers of spray lacquer and called them done (Figure **F**).

6. I'm familiar with gadgets and electronics and games, but I'm no coder. That being the case, I used RetroPie which is open-source software specifically for the playing of retro games on the Raspberry Pi. All that was needed was to download a RetroPie image of my choice, flash it to a microSD card using Etcher or other flashing software, plug it into the Pi, and boot it up.

7. I wanted to reserve two USB ports to be accessible from the exterior of the box so I ordered two USB extension cables and took measurements

F

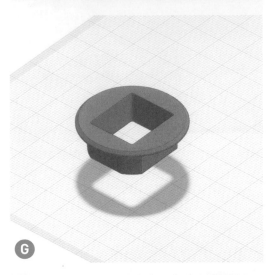

of the female ends. I used these measurements to custom design a pass-through piece in Tinkercad that would allow me to attach the USB cables from the interior and look nice and flush on the exterior (Figure **G**). I got the right fit with my second 3D-printed iteration.

8. The controls were all very plug and play. All I had to do was wire each button and joystick to its corresponding plug on the controller board and plug the board into the Pi via USB (Figure **H**). Then I configured the controls in RetroPie and everything worked smoothly.

9. The display is a salvaged 15.6" LCD screen from an old disused laptop (Figure **I**). I looked up the

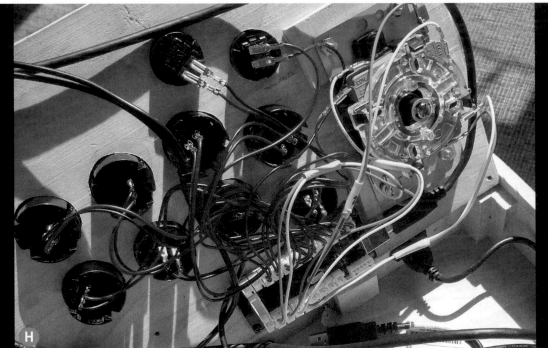

model number on eBay and found an appropriate LCD controller board. This would allow me to connect the screen (and repurposed speakers) to the Pi via a short HDMI cable.

10. For my sound solution, I repurposed some old desktop PC speakers. Breaking the speakers out of their casing required the use of a vise. However, once I got the speakers and electronics out I made the mistake of moving things around too much.

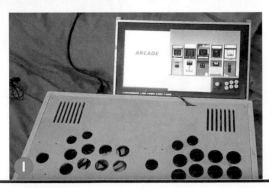

Wires that weren't supposed to move were jostled and inevitably broke. This required a lot of tiny resoldering and spiteful amounts of hot glue to remedy (Figure J).

11. The speakers sit in the base of the box, facing upward. I designed (again using Tinkercad) and 3D printed a pair of lifts to which the speakers could mount (Figure K). These allowed the speakers to be attached to the base, but be almost perfectly flush with the control face when it was laid into place over the top of the speakers.

12. To power all this, I wired the head of a 3-plug extension cord to an arcade power switch that passed through the back of the box and attached to a wall outlet via a standard PC power cable. The LCD controller board required a 12V, 1A power supply and I easily found one at a thrift store, but I had to modify it to fit inside the case (Figure L).

13. All the parts were done, all components tested and working. All that was left was to assemble the box. In the final stretch I found that the lid would not close over the joysticks as I had intended. I had already lowered the joysticks as far as I could with spacers and I found that no matter what angle I set the screen at, it would not close. I tried unscrewing the joystick ball tops from their posts and, voilà, the lid closed. There were mere millimeters of space between the screen and the bare posts, but this was enough for them to never touch.

I hadn't yet decided on a handle for the box so my solution was to add posts to the front of the box that the ball tops could attach to and be used as a carry handle. I did this without knowing if it would be a good idea because it was my only real option at this point. To my surprise it was ergonomic and quite comfortable to carry this way (Figure M).

I could not be more satisfied with the end result of my build. It works exactly as I had hoped and has already seen many rounds of *Street Fighter 2*. Odd as it may sound, the process of making this arcade case was a revelation. It was one of my first real projects and completing it was among the most rewarding experiences of my life. It gave me a new direction, new ideas, and new goals to strive for. The learning, problem solving, designing, and working in the shop felt intuitive and gave me a sense of being at home, which is a rare thing for me. I feel made to make and I can't wait to make more. ◗

The Gift of Giving

Written by the Little Free Libraries team

TIME REQUIRED: A Weekend

DIFFICULTY: Moderate

COST: $50–$100

MATERIALS

» Plywood, ¾", 4'×8' sheet
» Molding, pine quarter-round, ¾"×¾"×56"
» Acrylic glazing, clear, ⅛"×30"×30"
» Lumber, pine, 2×2 (nominal), 29" length
» Twin-wall polycarbonate glazing, 6mm, 22"×27"
» Polycarbonate U-channel for 6mm twin-wall glazing, 55" length
» Wood glue, waterproof
» Finish nails, 1¼" (12)
» Screws, pan-head, ¾" (8) for wood or MDF
» Screws, wood, 1¾" (54)
» Screws, wood grip, with neoprene washers #10×1½" (6) aka hex washer head roofing screws
» Hinges, exterior offset ("institutional") (2) for ¾" doorframe, with screws
» Door handle or pull (optional)
» Caulk, clear silicone
» Exterior wood finish or paint

TOOLS

» Eye and ear protection
» Work gloves
» Circular saw (optional)
» Jigsaw with wood blade and fine-tooth blade
» Straightedge
» Drill-driver
» Drill bits: ⅜", ¼", ³⁄₁₆", ⅛"
» Pilot-countersink bit
» Screwdriver bit
» ¼" nut driver bit
» Framing square
» Pencil compass
» Screwdriver (optional)
» ¼" manual nut driver (optional)
» Caulking gun

Larry Okrend

Like its little-library cousin, a "give box" offers a spot to share and borrow items of all sorts

A give box is one of the simplest and best examples of a community exchange. Anyone can put in anything they want (provided it fits in the box), and anyone can take out anything they want. You can be a giver or a getter, or both, anytime.

Give boxes seem to have origins in Europe, and at least one organization in the United States has formalized the idea to encourage free exchanges of food to help neighbors in need. But regardless of where a give box is planted, if it gets used, you can be sure it's helping someone out.

This version of a give box was adapted from several existing box designs and features an intentionally open interior to accommodate a wide variety of items. One side has two removable shelves; these are optional, and you can add more, if desired. The other side is wide open — a free space for standing up tall items or stacking others. It's even tall enough for hanging up adult-size shirts or jackets.

The roof, side windows, and door all let in natural light so passersby can easily get a hint of what's inside at a glance. The roofing is made from a single panel of clear polycarbonate twin-wall glazing, the same material that's used for the walls and roofs of greenhouses. You can buy it by the sheet at most home centers and some garden stores, and it cuts and drills much like acrylic glazing (which is what the windows and door front are made of, naturally). If you prefer not to have a clear roof, simply use a piece of plywood and cover it with any standard roofing material you like.

FIRST CUTS

Cut the side, back, and front panels and the door to size using a circular saw or jigsaw. Bevel the top edges of the front and back panels at 14 degrees to follow the roof slope. To mark these

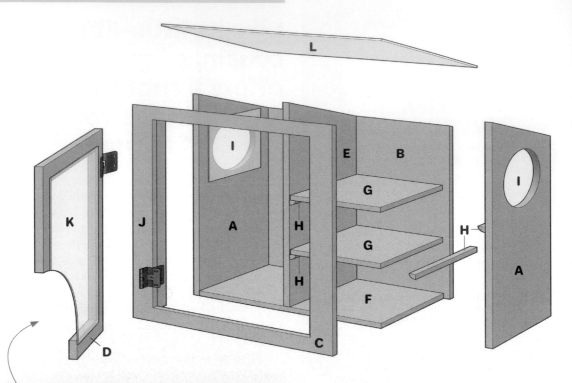

CUTTING LIST

PART | DIMENSIONS | QTY | MATERIAL

A – **Side panel** ¾"×13½"×32" (2) Plywood

B – **Back panel** ¾"×24"× 27¼" (1) Plywood

C – **Front panel** ¾"×24"×32¼" (1) Plywood

D – **Door** ¾"×20"×29" (1) Plywood

E – **Shelf divider** ¾"×13½"×31¼" (1) Plywood

F – **Base** ¾"×13½"×22 ½" (1) Plywood

G – **Shelf** ¾"×10¾"×13⅜" (2) Plywood

H – **Shelf support** ¾"×¾"×13½" (4)
Quarter-round molding

I – **Window glazing** ⅛"×10"×10" (2)
Acrylic glazing

J – **Hinge cleat** 1½"×1½"×29" (1) Pine

K – **Door glazing** ⅛"×16"×25" (1) Acrylic glazing

L – **Roofing** 6mm×22"×27" (1)
Polycarbonate twin-wall glazing

M – **U-channel** ¹¹⁄₁₆"×27" (2)
Polycarbonate U–channel

angled cuts on the side panels, measure down 4¾" from the top rear corner and make a mark along the back edge of the panel.

Cut the shelf divider and shelves to size.

DOOR AND FRONT-PANEL CUTOUTS

The door and the front panel each get a rectangular cutout made with a jigsaw. Make these cuts carefully, using a straightedge guide if desired, so the lines are nice and straight. You'll use the cutout material for the base of the box and the shelves.

Draw a 19"×28" rectangle on the backside of the front panel. Center the rectangle so it is 2" from each side edge and 2" from the top and bottom edges. Drill a ⅜"-diameter starter hole at each corner of the rectangle. Then use the jigsaw to cut between these starter holes to complete the cutout.

Repeat the same process to complete the door cutout, marking a rectangle that is 15"×24", centered so it is 2½" from each of the four edges (Figure Ⓐ).

Cut the box base to size using the plywood cutout from the front panel. Cut the shelves to size using the cutout material from the door.

CUT THE SIDE WINDOWS

Mark a vertical centerline on the inside face of each side panel, about 7" from the top edge. Measure up 22½" from the bottom edge of the panel and make a cross mark on the vertical centerline. This marks the centerpoint of the window cutout.

Draw an 8½"-diameter circle around the centerpoint using a pencil compass (Figure B). Drill a ⅜" starter hole inside the marked circle and complete the cutout with a jigsaw.

INSTALL THE SHELF SUPPORTS

The shelves can go on either side of the box. You can also position the shelves at any height you like. Cut the shelf supports to length from quarter-round molding using a jigsaw. Drill three pilot holes through each shelf support; fasten each with 1¼" finish nails and wood glue applied to one of the flat faces (Figure C).

ADD THE WINDOW GLAZING

Cut the windows to size using a jigsaw with a fine-tooth metal- or plastic-cutting blade. Drill a ³⁄₁₆" hole at each corner, 1½" from the side and top/bottom edges. The holes should be slightly larger than the threaded portion of the ¾" pan-head screws.

Remove the protective plastic from one face of each glazing piece and place the uncovered face over the window hole in the side panel. Fasten the glazing to the panel with pan-head screws (Figure D). Don't overtighten, as it might crack the glazing.

ASSEMBLE THE BOX

Mark layout lines for the shelf divider onto the inside faces of the front and back panels.

Apply glue to both side edges of the base. Fit the side panels over the base so they are flush with the base at the front and rear. Fasten through each side and into the base with three 1¾" wood screws. Repeat with the rear edges of the side panels and the base.

Make three evenly spaced marks (from top to bottom) on the backside of the back panel, 12" from either side edge. Apply glue to the rear edge of the shelf divider. Place the divider on the base and against the back panel, aligning the divider between its layout marks on the back panel. Drive three 1¾" screws through the back panel and into the divider, placing a screw at each of your marks on the back panel.

Glue the front edges of the side panels and base. Apply a line of glue between the marks on the inside face of the front panel. Fit the front

Christopher R Mills, Larry Okrend

E

F

G

H

I

panel over the box so all pieces are flush on the outside and top, and the divider is in between its layout marks (Figure **E**). Fasten the front panel with four 1¾" screws into the side panels and base. Also drive two screws into the divider through the panel, about 1" each from the top or bottom of the front panel.

ADD THE HINGE CLEAT

Cut the hinge cleat to length and glue to the inside of the front panel so it is flush with the edge of the door opening. Fasten with three 1¾" screws.

PREPARE THE ROOFING

Cut the polycarbonate roofing to size using a jigsaw with the same blade used for the acrylic glazing. (You can also use a circular saw with a hollow-ground panel blade with 10 to 12 teeth per inch.) Note that the lines in the panel will run parallel to the sides of the box when the roofing is installed.

Cut the two pieces of polycarbonate U-channel to length with a jigsaw. Fit pieces of U-channel over the front and rear edges of the roofing panel so they're flush at both sides. The U-channel keeps bugs, rain, and snow out of the hollow cells of the roofing material.

Drill three ⅛" holes through the bottom U-channel, aligning each hole with one of the hollow cells in the roofing. These are weep holes that will allow condensation to drain from the cells.

INSTALL THE ROOFING

Set the roofing panel on the assembled box so it overhangs about 4" at the front of the box and equally at both sides. Drill a ¼" pilot hole through the roofing near each corner of the box, about 1½" from the front and rear faces. The holes are slightly larger than the screw shank to provide room for expansion of the plastic roofing.

Be careful to drill just through the roofing and no more than about ⅛" into the plywood; if you drill too deep, the screws won't hold. Also make sure the holes go through the hollow part of a cell and not into a cell wall.

Drill two more pilot holes over the shelf divider, about 1½" from the front and rear of the box.

Fasten the roofing to the side panels and shelf

Larry Okrend

divider with six #10×1½" wood grip screws driven through the pilot holes. Drive the screws so they are snug to the panel and the neoprene washer under the screw head is slightly compressed (Figure **F**). Be careful not to drive too far, which can crack the panel. (You might want to use a ¼" manual nut driver here for better control.)

ADD THE DOOR GLAZING

Cut the door glazing to size using a jigsaw, as with the window glazing. Mark and drill three pilot holes along each edge of the glazing, ¼" from the outside edge. Place one hole about 1½" from each corner and one hole centered in between.

Center the glazing over the door so it overlaps the door cutout by ½" on all sides (Figure **G**). Fasten the glazing to the door with twelve ¾" pan-head screws, again being careful not to overtighten the screws so you don't crack the acrylic glazing.

HANG THE DOOR

Mount the door hinges to the backside of the door using the provided screws.

Mount the door to the front panel, screwing into the side edge of the door cutout with 1¾" wood screws (Figure **H**). The door should overlap the cutout equally at the top and bottom.

FINISH THE PROJECT

Remove the door from the box, then remove the hinges and the door glazing and window glazing. Keep the hardware in a safe place for re-installation later.

Apply an exterior finish to the outside of the box and to all surfaces of the door. If you'd like to paint your project, you can create the appearance of a seamless box by covering the plywood edges with auto body filler before painting.

After the finish has fully cured, reinstall the door glazing, then mount the door hinges and hang the door. Install a door handle or pull, as desired. Add the shelves by setting them on the shelf supports.

Finally, seal around the window and door cutouts with a fine bead of clear silicone caulk (Figure **I**). Let the caulk cure overnight before installing or using the box. ⊘

Light It Up

The urge to dive into some gripping literature can occur any time of day or night. Help your neighbors find just the right title with this simple solar-light mod for your Little Free Library, as devised by Kim and Bill Anderson from Waterloo, Ontario.

1. Remove the lawn post from a narrow-cylinder solar lawn lamp or solar rope light.
2. Using a hole-saw with the same diameter as the shaft on the solar panel, drill a hole near the interior roofline of your box.
3. Insert the lamp with a ring of silicone sealant so that the solar array remains exposed to outside light. Align the lamp or lights inside.

Happy reading!

Embroidery Animation

Make your needlework come alive in awesome little movies!

Written and photographed by Alexis Sugden

ALEXIS SUGDEN
is an animator,
comic creator,
and embroiderer.
Currently an
animation director
on Disney Junior's
T.O.T.S., she's an
Australian living in
Vancouver, British
Columbia. linktr.ee/
alexisdraws

About a year and a half ago, I started getting into embroidery. Once I got the hang of it, I realized that embroidery is just like drawing, with thread. And if I can draw, I can animate!

I've done a few different embroidery animations, including a joyful (and nonbinary) dancing Office Bat that went viral last year. This took 24 separate embroideries and about 3 months of my spare time to make. Of course I am working on more. You can see my embroidery animations at vimeo.com/alexissugden.

If you can animate and embroider, or you're curious about the process, this how-to is for you.

TIME REQUIRED: A Month of Evenings

DIFFICULTY: Intermediate

COST: $10–$30

MATERIALS
» **Embroidery thread**
» **Fabric**

TOOLS
» **Needle** and a thimble doesn't hurt!
» **Scissors**
» **Sewing pins**
» **Embroidery hoop**
» **Pencil**
» **Lightbox**
» **Computer or tablet** with animation software of your choice; see Step 2 below.
» **Printer**
» **Iron and ironing board**
» **Camera and tripod**

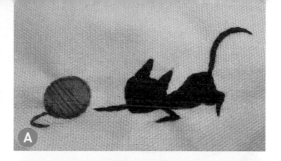

1. DESIGN AND TEST

Design your character and do a little test to see what size works well, what colors of thread, what fabric, etc. It's easiest to figure that out before you start, rather than changing your mind when you're 10 embroideries deep.

I recommend starting with a very simple character, using only a few colors. You can sew everything quite small, as long as it's captured with a good camera. You can also make things easier by just stitching the outline of a character, or by creating the fills with long stitches rather than many small stitches that will take a long time. Remember that whatever you plan to sew, you will sew many, many times, so you want to make it as easy for yourself as possible.

Figures **A** through **D** show some color tests I did for my *Embroidery Cat* animation. Figure **E** is the design I used for my *Office Bat Boogie* animation. As you can see, the cat design is much simpler. Both animations took me the same amount of time, but I was able to stitch twice as many frames with the simpler, smaller cat design.

When choosing fabric, look for one that won't warp too much when you take it out of the embroidery hoop. I found that felt, for example, warped a lot, but cotton with a thick weave worked better.

(F)

(G)

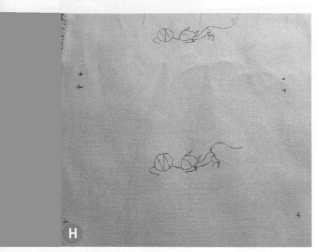

(H)

2. ANIMATE!

In order to create embroidery animation, you need to animate. I prefer to animate in 2D for my embroidery animations, but you could also animate in 3D. Personally, I think this would give a rotoscoped feeling, so I've avoided using 3D for this reason.

There are several different programs you could use to create your animations. On your computer you could use TVPaint (my personal favorite), Adobe Animate, Toon Boom, or Photoshop; if you're on an iPad or tablet, you might use ProCreate. Pencil2D on the computer and RoughAnimator on the iPad are free animation software, if you don't plan on animating often, and don't want to sink money into expensive software.

If this is your first time doing embroidery animation, I recommend creating a short loop, so you can get the most out of your embroideries. For my first embroidery animation, I made an 11-frame loop of a jumping *Frog*, to keep things easy for myself.

I also recommend animating on threes or fours, to help keep the number of embroideries down. The tests I've done so far have been on twos, but my next, longer, project will be on threes.

ANIMATION TERMINOLOGY: People usually animate at 24 frames per second. Animating *on twos* means that each drawing is held for 2 frames, therefore you would need 12 drawings per second. Animating *on threes* means holding each drawing for 3 frames, *on fours* holds each drawing for 4 frames, and so on.

Make sure to clean up your animation so that it has clear, defined lines and isn't too loose or scribbly. You'll be using this as a guide to your sewing, so you want it to be as clean as possible. Another thing to do now is to track the direction of stitches. For example, when animating this cat and ball, I drew guidelines for the stitch directions so that the ball would feel believable when rolling. The lines on the ball in Figure (F) are a guide for the direction of the stitches in Figure (G).

3. PRINT AND TRACE

Once your animation is clean and complete, print it all out.

Using a lightbox and a pencil, trace the drawings onto your fabric. I don't hem my fabric, so I make sure there's enough room for the fabric to fray without destroying the space I'm using.

I recommend drawing registration points for each frame (Figure (H)) so that your photographs of your embroideries will be easier to line up later, as well as numbering each frame.

> **TIP:** A lightbox or small light table is an easy DIY project; see makezine.com/projects/photo-lamp-and-lightbox for an example.

4. EMBROIDERY TIME!

This step is the most time consuming. Because the embroideries need to follow the drawings as closely as possible, I like to stitch the outlines before doing the fills. Otherwise, I would have trouble finding the exact lines after doing the fills.

For example, if I was just doing one embroidery

of Office Bat, I would do the fills first, then the lines. But in the animation, I made sure to do the glasses and face before the fills, so that I didn't lose those important lines.

Make sure to stay consistent with the order in which you stitch everything. For Office Bat, I would first stitch their pants, then shirt, tie, glasses and face, and finally head, body, and wings. Again, with each step I was stitching the outlines before doing the fills.

Also keep the direction of the fill stitches consistent. For example, if you stitch the fill of the tie in one direction in one frame, make sure it's the same in all frames, otherwise the animation will get too jittery.

Figures **I** and **J** are work-in-progress shots, as I'm sewing over the pencil lines.

5. IRON YOUR EMBROIDERY

Try to iron out the hoop crease as much as you can (Figure **K**). When working on a piece, I try to release my fabric from the embroidery hoop whenever I'm not working on it, so that it's easier to iron out later.

6. CAPTURE YOUR EMBROIDERY

At first I tried scanning my embroideries, but I found that photography works much better. I'm not a very good photographer, so I got some help from my friend Will Robson (@_willrobson on Instagram).

Make sure you're using a tripod so you can take every photograph from the exact same distance. Don't worry about lining them up exactly, just make sure the registration points are in the photographs so you can line everything up on the computer.

7. ASSEMBLE YOUR IMAGES

Bring the photographs onto your computer, it's time to assemble your animation! For this, I used Photoshop, purely because I've been using it for so long that I'm very comfortable with it, and find it easy to use.

Make a file with the dimensions you want to export at, and size your images to fit. When scaling your images, make sure you scale them all by the same amount.

You can use your registration points to help line up your images. I also keep a layer in there with the original hand-drawn animation so that I can line up the embroidery with the original. Your fabric might have warped a little while sewing, so things won't always align perfectly. I like to make sure the feet are working, and then the rest seems to work just fine.

8. EXPORT AND ENJOY

Show your animation to everyone (Figure **L**) and make them pay you compliments, because this process is a lot of work! ●

Lex Sugden 🌱 @AlexisDraws · Aug 19, 2019
At last! Every frame hand embroidered to bring you the joy of this non binary office bat, just dancing to the beat of their own drum <3

1.2M views   0:02 / 0:04

○ 759   ⟲ 23.7K   ♡ 85.7K

Solar Flares and EMP

Written by Forrest M. Mims III

Prepare for solar eruptions and nuclear weapons with a DIY Faraday pouch to shield electronic gear

Our dependence on electricity has made the entire world vulnerable to an existential threat. That threat is what might occur should there be a prolonged cutoff of electrical power to cities and even countries.

At least two kinds of events might cause a widespread loss of electrical power: a major coronal mass ejection (CME) from the sun directed toward Earth and the electromagnetic pulse (EMP) from the detonation of a nuclear explosion above the atmosphere. These events have the potential to shut down regional or even international electrical grids for days, months, or even years if high-voltage transformers at power plants and substations are destroyed by the huge electrical currents that a CME or EMP event can induce in high-voltage transmission lines.

The EMP from a nuclear explosion includes a secondary threat, for its nanosecond rise-time has the potential of damaging many kinds of electronic devices not connected to the grid.

SOLAR FLARES

On September 1, 1859, British astronomer Richard Carrington was sketching a cluster of sunspots viewed through his solar telescope when he saw what he described as "two patches of intensely bright and white light." Fellow astronomer Richard Hodgson also observed the phenomenon.

They had observed a massive solar flare accompanied by a CME that propelled an immensely powerful burst of energy directly toward Earth (Figure A). Less than a day later, the CME arrived and triggered brilliant, worldwide auroras. Telegraph operators reported large sparks in their equipment, and some received electrical shocks.

TIME REQUIRED:
30–60 Minutes

DIFFICULTY:
Easy

COST:
$25–$50

MATERIALS
» **Conductive fabric** intended for EM shielding. I used Mission Darkness fabric from mosequipment.com.
» **Conductive tape** such as TitanRF Faraday tape, also from mosequipment.com
» **Static protection foil bag** such as SCS Dri-Shield 3400, available at digikey.com
» **Velcro tape, self-adhesive**

TOOLS
» Scissors

FORREST M. MIMS III an amateur scientist and Rolex Award winner, was named by *Discover* magazine as one of the "50 Best Brains in Science." His books have sold more than 7 million copies. forrestmims.org

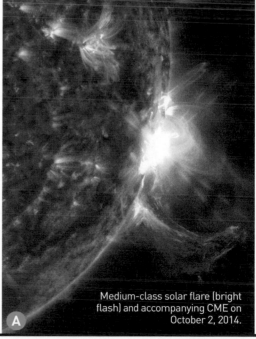

Medium-class solar flare (bright flash) and accompanying CME on October 2, 2014.

Adobe Stock-Kittiphat, NASA's Solar Dynamics Laboratory

A

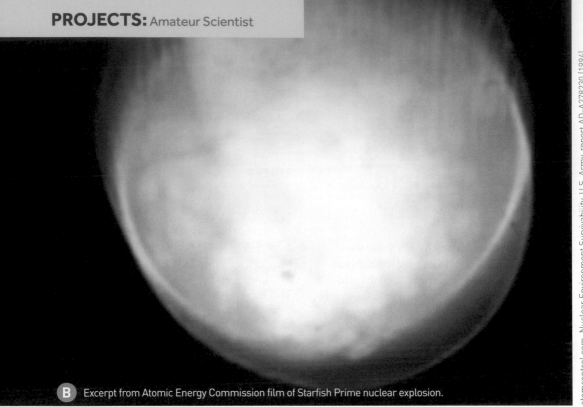

atomcentral.com, Nuclear Environment Survivability, U.S. Army, report AD-A278230 (1994)

B Excerpt from Atomic Energy Commission film of Starfish Prime nuclear explosion.

This historic phenomenon, which was named the Carrington Event, was the first known visual observation of a solar flare. The telegraph disruptions occurred because the wires strung from poles across the countryside served as antennas that received the oncoming energy and transmitted it as an electrical current.

What might occur today if a Carrington-scale solar storm occurred? A hint occurred on March 13, 1989, when a much smaller CME created a geomagnetic disturbance (GMD) that disrupted satellites, caused a 9-hour power failure across Quebec, Canada, and destroyed a multi-million-

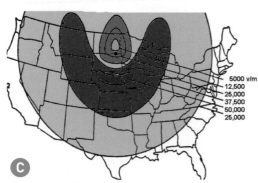

5000 v/m
12,500
25,000
37,500
50,000
25,000

C

Estimated EMP in volts/meter, from a nuclear weapon detonated high above the United States.

dollar transformer at New Jersey's Salem Nuclear Power Plant. A Carrington-scale CME has the potential of shutting down electrical grids around the world, and that almost happened on July 23, 2012, when the most powerful CME ever monitored passed close to Earth.

Should a potentially damaging CME event occur, the National Oceanic and Atmospheric Administration (NOAA) will post advisories on its Space Weather Prediction Center website (swpc.noaa.gov) half a day or so before it arrives.

NUCLEAR EMP

A nuclear explosion generates a powerful electromagnetic pulse having an electric field of thousands of volts per meter (the potential difference between two points 1 meter apart). I've measured 2,000 volts/meter directly under high-voltage transmission lines. A nuclear EMP would create a much higher voltage in an ultra-fast pulse over many thousands of square miles.

On July 9, 1962, the Starfish Prime nuclear bomb was detonated 250 miles over Johnston Atoll, 826 miles from Honolulu. Figure **B** is a frame from an Atomic Energy Commission film of the event, provided by atomcentral.com; you can watch it at youtube.com/watch?v=XoXtUkcFmG4. An EMP of about 5,600 volts/meter from the blast disabled at least six satellites and a few hundred streetlights in Hawaii.

In his report "Soviet Test 184: The 1962 Soviet Nuclear EMP Tests over Kazakhstan," Jerry Emanuelson describes the damages caused by the estimated 10,000 volts/meter EMP from a Soviet nuclear test 180 miles over Kazakhstan on October 22, 1962. Facilities damaged included a power plant, diesel generators, a radar system, and lengthy buried and above-ground cables.

Much more powerful EMP bombs have been developed by China, Russia, and the United States. North Korea has claimed it could shut down the U.S. power grid with an EMP bomb. James Stavridis, a retired U.S. Navy admiral and former NATO commander, is deeply concerned that North Korea might someday attempt an EMP attack against the United States. Figure Ⓒ shows the estimated EMP in volts/meter from a bomb detonated high over the central U.S.

Lightning is a natural form of EMP, and engineers have developed many ways to protect transformers from lightning strikes. While the duration of the secondary phases of a nuclear EMP resembles lightning, there is concern that older transformers are at serious risk. Then there's the initial phase of the pulse, which has a rise time measured in nanoseconds with maximum power between 100kHz and 1GHz. This pulse can be fatal to semiconductor components such as CMOS integrated circuits not connected to the grid.

EMP VS. PERSONAL ELECTRONICS

A solar CME and a nuclear EMP both pose a potential threat to most electronic systems connected to the grid. Personal electronics not connected to the grid or cables will probably be unaffected by a CME but might be damaged or even fried by an EMP. Small devices like electronic watches, calculators, radios, and phones might be safe, but no one knows.

A Faraday shield that blocks electromagnetic radiation is the best way to reduce the effect of a nuclear EMP on personal electronics. The ideal Faraday shield is a fully enclosed metal enclosure with no unshielded external wires or cables that would function as antennas and thereby couple EMP into the enclosure.

Many kinds of DIY Faraday shields for protecting personal electronics can be found online. Among the best sites is disasterpreparer.com by Arthur Bradley, a former NASA engineer who has a doctorate in electrical engineering.

Bradley knows his stuff. For example, many sites claim a Faraday shield must be grounded. Bradley denies this, and he's right. Bradley has used sophisticated instrumentation to test the most common Faraday cages described online. He has given the results in YouTube videos accessible on his website and in *Disaster Preparedness for EMP Attacks and Solar Storms*, a book available in print or as a Kindle ebook from Amazon.

The effectiveness of Faraday cages in reducing an EMP is given in the following chart in decibels.

Shielding in decibels (dB)	EMP reduction factor	Peak voltage 25 kV/m EMP	Peak voltage 50 kV/m EMP
10	3.2	7,900	15,800
20	10	2,500	5,000
30	32	790	1,580
40	100	250	500
50	316	79	158
60	1,000	25	50

D A DIY Faraday bag — an antistatic bag with an inner conductive fabric lining — can provide up to 50dB protection for small electronics.

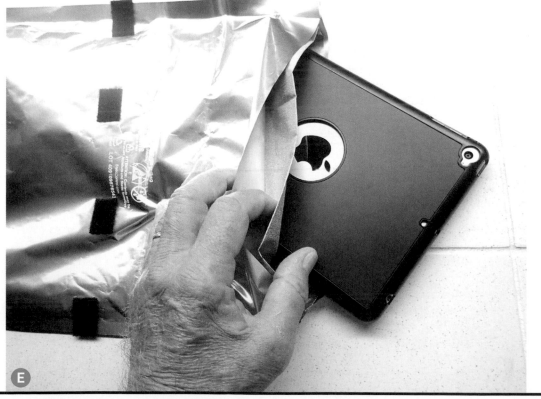

E

Bradley recommends providing shielding with at least 50dB attenuation. His YouTube videos include tests of the EMP reduction provided by a simple Faraday cage made by wrapping a cardboard box with aluminum foil. If the foil provides a tight shield when the box is closed, his tests shows that it will provide a shielding of >50dB across most of the EMP spectrum. Sealing the foil-lined lid with aluminum tape helps. His YouTube tests include Faraday shields made from conductive fabric, static bags, ammo boxes, and trash cans.

Shielded static bags are among the simplest and easiest to use Faraday shields. Bradley's tests show that SCS Dri-Shield 3400 moisture barrier bags are the best. You can buy these from disasterpreparer.com and digikey.com.

Figures **D** through **F** show an EMP shield for a tablet that I quickly made from a DIY conductive fabric pouch inserted inside an SCS 3400 bag with Velcro closures I added. Conductive fabric is available from many sources, including Bradley's site. I used Mission Darkness fabric sealed with their tough TitanRF Faraday Tape (mosequipment.com). This simple Faraday cage provides excellent shielding and can be used for laptops, radios, phones, and other small devices potentially vulnerable to an EMP.

Mission Darkness has a free app that provides an estimate of the EMP blocking capability of Faraday shields by using Wi-Fi and phone signals as surrogates for EMP (Figure **G**). Activate the app on a phone, insert the phone in the enclosure for 30 seconds, and then remove it to see the results.

LEARNING MORE

You can find considerable information online about the hazards presented by a major CME and an EMP attack. A good brief summary is "Report to the Commission to Assess the Threat to the United States from Electromagnetic Pulse (EMP) Attack: Risk-Based National Infrastructure Protection Priorities for EMP and Solar Storms" by George H. Baker (July 2017).

Finally, if an EMP or GMD event shuts down power to law enforcement, water and sewer systems, gas stations, grocery stores, and hospitals, the least of your concerns might be

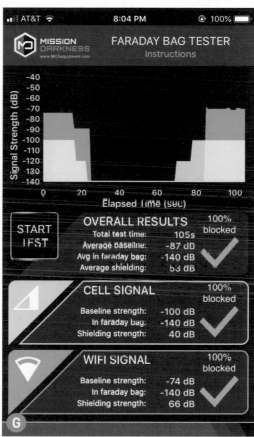

Mission Darkness Faraday shield testing app.

protecting personal electronics. So be sure to check out the survival recommendations at redcross.org and private sites. ●

Congressional EMP commission reports: firstempcommission.org

1+2+3 Mid-Century Sunburst Mirror

Written and photographed by Mara Capron

Use zip ties and hot glue to add retro flair to everyday things

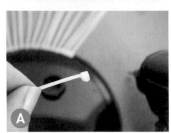
A

B

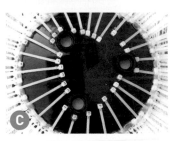
C

D

E

This modern mid-century sunburst mirror is by far my favorite DIY I've ever made!
It's easy *and* cheap to make; it takes about an hour and can cost $10 or less. I love it so much!

1. Prep your mirror

Plug in your glue gun. Read the instructions for how long it needs to warm up (mine takes about 5 minutes). Clean your mirror of any dust. Then tape around the interior of the mirror and place a trimmed piece of cardboard to protect the mirror when you spray paint it.

Next, plan out your zip tie placement before you start gluing. You can follow my design or create your own.

2. Glue the zip ties

Squirt a small drop of hot glue onto the head of the zip tie (Figure Ⓐ) and press it down on the back of the mirror.

NOTE: I found it difficult to hold the glue gun and the zip ties and keep the mirror steady at the same time since the glue had already dried by the time I put the glue gun down and picked up a zip tie. So a helper with a second pair of hands is useful. Alternatively, you can keep your glue gun standing up to the side and squirt out the glue without picking it up.

Row 1: Place the smaller zip ties all around the inner lip on the back of the mirror (Figure Ⓑ). After you finish, clean up all the little glue hairs left behind.

Row 2: Place the longer zip ties right behind the first row, every 3 or so.

Row 3: Place longer zip ties further toward the interior of the back of the mirror, so that their tips will lie in between the first and second rows (Figure Ⓒ).

TIP: After placing all the zip ties, I added more hot glue over rows 1 and 2 and where row 3 intersected at the edge of the mirror. This ended up being a good idea since a few weeks later one zip tie from row 3 came loose, but didn't fall off since I had glued all the rows together at the edge. I just put a dab of hot glue on the loose one, and it was as good as new.

TIME REQUIRED:
An Hour
DIFFICULTY:
Easy
COST:
$10–$20

MATERIALS
» **Makeup mirror, 5" or 13cm round** around $4 on AliExpress or $9 at Walmart
» **Zip ties: 4"/10cm (110 pieces) and 8"/20cm (60 pieces)**
» **Acrylic spray paint, gold**
» **Hot glue gun with glue sticks**

MARA CAPRON is opening a bed & breakfast with her husband and working on multiple DIY projects to make the rooms trendy without the expense.

Remove any little cobwebby glue before spray painting.

3. Add some gold

Now you can spray paint the cardboard and zip ties. Be sure to work in a well ventilated space.

TIP: Cover your workspace with cardboard or a plastic bag, so as not to get paint all over.

Following the directions on the can, spray the back of the mirror and allow to dry. Then flip the mirror over and spray the front (Figure Ⓓ). I used one coat on the back and three on the front, but it's up to you.

Once your paint is dry, remove the protective cardboard and tape, and admire your sunburst!

Make Things Bling

Of course you could hang this on your wall, but I thought it would be more photogenic for me to bling out my Vespa (Figure Ⓔ). I don't wear gold jewelry, but now I want to spray paint everything gold! ◐

Growing
LEATHER

Use kombucha to make this versatile vegan textile in your kitchen

Written by Christine Knobel

TIME REQUIRED
Growing: 2–3 Weeks
Harvesting: 5–10 Minutes
Drying: 1 Week/layer

DIFFICULTY
Easy; patience is the most demanding skill.

COST
$16–$20

CHRISTINE KNOBEL
works as a freelance designer in the Northern California Bay Area. Her focus is in fashion and sustainable design. She continues to explore kombucha as well as other materials and methods to decrease waste in cosplay with the hope that it can also be carried over into the fashion industry.

ALL IN THE NAME: Kombucha, the refreshing fizzy fermented tea beverage, shares the same name with the kombucha that can be used for truly vegan ecofriendly leather. Kombucha in this article refers to that very leather.

A

Kombucha leather differs from traditional vegan leather because it is not synthetically based like plastic. It is instead biologically based, from bacteria and yeast. Due to its biodegradable nature, it's great for shorter-life projects like wallets, shoes, clothing, costumes, and cosplay. For anything that doesn't need to last decades, kombucha will do. As a bonus, at the end of the product's lifecycle it can be thrown in the compost.

This kombucha textile begins with sweet tea, either black or green, and a bit of kombucha SCOBY ("Symbiotic Colony of Bacteria and Yeast") added after the tea is cooled. The process takes about 2 weeks to form.

The textile is the result of the bacteria in the SCOBY creating a *pellicle*. The pellicle is the skin or film that's formed by the SCOBY, allowing it to float and giving the aerobic bacteria access to consume oxygen (Figure **A**). The yeast in the SCOBY consumes the sugar and ferments the tea into the kombucha drink.

The pellicle is often referred to as *bacterial cellulose* in scientific papers and is the most pure form of cellulose on the planet. It is 100% biodegradable.

A FEW SIMPLE INGREDIENTS are needed for brewing kombucha. This recipe can be multiplied. I typically brew 4 to 8 gallons at a time depending on my project.
* **1 gallon water**
* **1 cup of sugar** White or table sugar is ideal, but feel free to experiment with brown sugars and pasteurized honey.
* **½ ounce of loose leaf tea, black or black/green blend** or 6–7 black tea bags
* **½ cup of starter kombucha** i.e. already brewed kombucha
* **1–2 small pieces of SCOBY, or 2 bottles of kombucha** Time and results are the same.

A FEW SIMPLE TOOLS are needed for production:
* **Stovetop** or other way to brew a minimum of 1 gallon of tea at a time
* **Large pot and spoon** to brew sweet tea
* **Container(s)** for growing kombucha, ideally plastic or glass
* **Container** to keep your mother SCOBY happy, ideally glass

Note that kombucha SCOBY will grow in the shape of its container, no matter the profile: circular, square, etc. I used under-the-bed storage containers that are about 32"×12" for growing large sheets. I brew enough for two to four containers at a time and fill these containers with 2 gallons of sweet tea each.

Directions

1. Bring the water to a boil and dissolve the sugar.
2. Turn off the stove and add the tea (Figure **B**). Let it steep for 15 minutes, minimum. You can also brew it in the evening and let it cool, covered, overnight. Remove the tea bags or loose leaf tea after steeping.
3. Once the tea is room temperature, add it to the brewing container (Figure **C**). This can be glass or plastic. Make sure there is 4"–5" of sweet tea in the container for growing kombucha sheets.
4. Add the starter kombucha and 1–2 pieces of SCOBY, or two bottles of bottled kombucha.
5. Cover the container with a thin sheet of fabric and rubber band, and/or loose lid (fruit flies love kombucha and the fabric or lid will keep them from the tea). Make sure there is some air flow to the kombucha; don't seal a lid shut on the container.
6. Leave the kombucha undisturbed in a dark, warm space for optimal growth. To expedite growth, keep the container in an 80°F–95°F environment. This will yield about ½" sheet of kombucha in 2 weeks. The pellicle will also grow in a cooler environment, but it will take longer to grow a viable thickness.
7. Once the kombucha pellicle is about ½" thick, it is time to harvest (Figure **D**)!

Drying It Out

The drying process can take as long as the growing process, but it requires more handling. The drying time will depend on the environment, temperature and humidity, and the quantity of layers desired. There will be considerable shrinkage in the thickness of the pellicle as it dries. Expect up to 90% of its thickness to be lost to evaporation.

No matter what your desired result is, the most important thing is to layer multiple sheets of kombucha as it dries (Figure **E**) — this will increase its strength and flexibility.

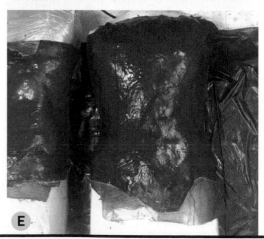

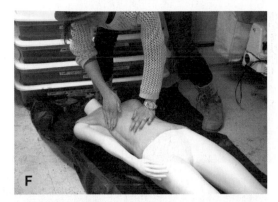

F

G

H

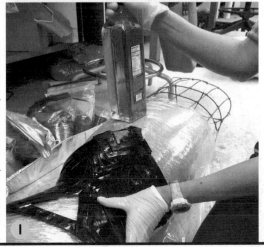

I

When a sheet of kombucha SCOBY dries, the nano-crystalline structure of its cellulose will align in a single direction. Each sheet will align independently, even if fused with other sheets. Layering these nano-crystalline structures in multiple directions by fusing multiple sheets increases the strength of the leather and reduces the chance of tearing and cracking when dried.

Sheets can be layered while wet, or a wet sheet can be added to a dry or almost dry sheet. This is how kombucha sheets fuse. One sheet can be grown at a time. A second sheet can be growing while the first sheet is drying. The fusing result will be the same whether fused wet-to-wet or wet-to-dry, the only difference will be the time. It's faster to allow the first sheet to dry most of the way before adding the next layer.

Plan a minimum of four layers of kombucha to create a truly viable piece of vegan leather. This way its strength and flexibility will be increased as well as its dry thickness. You should also plan for some shrinkage in the length and height as the kombucha dries, but with daily massaging (Figure F) this will be minimal.

You can dry kombucha on a wet mold to create a three-dimensional shape (Figure G), or flat like traditional textile sheets (Figure H). The kombucha will pick up the small details of the surface that is dries upon. This makes creating an embossed sheet or form almost effortless.

Apply a thin layer of olive oil on the drying surface (either a mold or a flat plastic sheet) to prevent the kombucha from sticking when dried (Figure I). Olive oil will also make removal a breeze.

Remember to massage the kombucha sheets daily for a smooth, even texture when dried. If no massaging is done during the drying process the dried piece of kombucha will prune like a raisin and resemble wrinkly skin. Experiment with massaging levels to create different textures.

Working With Dried Sheets

Once dry, kombucha will behave much like leather. It can be cut, stitched, glued, painted, and dyed. It cuts easily with scissors in any direction and will not fray. Do be mindful that the scissors are stainless steel and that the kombucha is fully dry. If not, the scissors may rust.

• • • Each sheet of kombucha is likely to have imperfections like thin spots or holes where gas was released during the fermentation process. This can be solved two ways: either by patching the holes with other pieces of kombucha, or by creating sheets of blended kombucha.

There is evidence that blended sheets of kombucha can still dry with their nano-crystalline structures aligning in a single direction. So an imperfect sheet of kombucha can be harvested, blended into tiny pieces, and laid out in a sheet to dry. Once the first blended sheet is almost dry, you can add the next blended layer on top, or make a sandwich of full sheets and blended layers combined. Feel free to experiment to create the ideal textile for your desired project.

Sewing

Kombucha can be sewn beautifully by machine or by hand, just the same as leather. Use a leather needle, Teflon or walking foot, and a longer stitch length (4–6) when using the sewing machine. Again, work with fully dried kombucha when sewing. Otherwise the feed dogs and/or bobbin casing can rust. A leather punch can be used to create the holes for hand stitching.

Fusing

Kombucha has this incredible ability to "heal" by fusing wet pieces over dry or almost dry sheets. The kombucha product is easy to repair if any tears, cracks, or holes occur. Figures **J** and **K** show a cracked piece before and after repair.

This ability also lends itself to attaching two or more pieces together. Seams can be fused. This is the ideal method if the desire is a 100% biodegradable product, but it also takes the longest time to complete. If there is a need for haste, rubber or contact cement can be used at the seams (this portion will need to be cut away before adding the rest of the kombucha product to the compost heap).

Dyeing and Painting

Being the most pure form of cellulose, kombucha is an ideal candidate for dyeing. It can be dyed when first harvested or after it is dried. When considering the weight for dye quantity calculations, use the weight of a dry sheet of kombucha. (Remember, the freshly harvested pellicles are mostly water.) Using natural dyes (Figures **L** and **M**) will create a stiffer sheet of kombucha when dry, while a fiber reactive (procion) dye will increase the flexibility of the sheet. Experiment with both to create the right

J

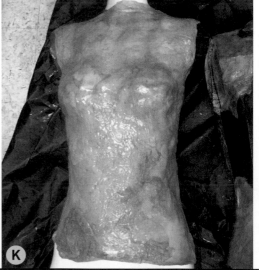

K

Jenni Cadieux, Christine Knobel, Hannah Kent, Heroic Images

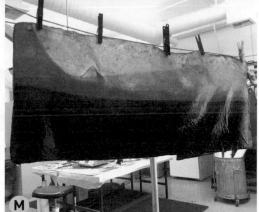

color and texture for your kombucha product.

Paints and pigments can also be used to color the surface of the kombucha (Figure N). Cake paints work well and keeps the product fully biodegradable. Acrylic paints can also be used. Some increase in the material's flexibility has been observed when using acrylics. As the kombucha ages, the paints darken in color. These can be touched up as needed.

Care

Humidity levels will have a great impact on the life of the kombucha product. Lower humidity will lead to more stiffness, while higher humidity will increase flexibility. Use the techniques above to create a kombucha textile to suit your climate for the longest lifespan.

Kombucha resembles human skin when it dries without paint, pigment, or dye. Like human skin, it responds positively to lotions and oils. If traveling to a lower humidity climate, give it some lotion or body oil. Or use lotions to increase the shine. If cracks or tears do occur, heal it with a fresh piece of kombucha. The resulting patch can be sanded, once fully dry and fused, to create an even surface.

Kombucha may not be as durable as leather, but it can be completely composted at the end of its lifecycle. Kombucha vegan leather has a smaller environmental impact than PVC vinyl and actual leather. It is fun to grow right in your kitchen or on your back porch, and you can nurture a garden with it by adding it to the compost at the end of its use. Pretty neat, right? ⊘

Full finished vegan leather Wonder Woman costume.

Get in GEAR

Learn the options for transmitting movement in your mechanical projects

Written by Brian Bunnell and Samer Najia

You can find this article and much more useful information in our new book *Make: Mechanical Engineering for Makers*, available now at all major booksellers and at makezine.com/go/mechanical-engineering-for-makers.

Gears come in a wide variety of shapes and sizes, and are used in nearly every mechanical system. Here, we introduce you to some of the most common gear types that you may come into contact with as a maker.

● **SPUR GEARS** are the simplest of gears, and are what most people envision when thinking of gears. They are wheels or cylinders with teeth arranged radially outward from the center (Figure Ⓐ). When working as a system (e.g. with multiple spur gears meshed together), the shafts of spur gears must be parallel to each other. Spur gears work well at lower speeds, but are noisy at high speeds because of the gear teeth impacting each other as they engage.

● **HELICAL GEARS** are wheels or cylinders with the teeth cut at an angle relative to the axis of the gear (Figure Ⓑ). The teeth are curved along a helix shape (hence the name helical gear; they are sometimes called *herringbone gears*). Due to this helical profile, they engage more gradually than spur gears, thus providing smoother and quieter running gears. Helical gears can be arranged with their shafts parallel or at various angles relative to each other, depending on the teeth configuration. When in operation, helical gears tend to have higher friction due to the "sliding" of the gear teeth against each other. Also, helical gears have an *axial load* or *thrust load* on the gear shaft, since the gear teeth engage along a curve instead of straight on; in other words, the gear tends to "screw in or out" as it rotates. Double helix gears eliminate this issue by having a double set of teeth oriented in opposite directions. The axial thrust developed by one side of the gear is negated by the axial thrust generated by the other side in the opposite direction. Double helix gears are difficult to make and are very

Ⓐ Spur gear set

Ⓑ Helical gear set

Brian Bunnell

expensive, but they are necessary in certain instances.

- **BEVEL GEARS** are wheels or cylinders with teeth arranged around a cylindrical cone with the tip lopped off. When two bevel gears with the same diameter and number of teeth are working together in a system, they are known as *miter gears*. Miter gears only change the gear shaft axis angle or axis of rotation; there is no change in rotational speed relative to the input or output gear. Similar to spur gears, bevel gears run well at lower speeds, but they get noisy at higher speeds. Engineers have combatted this issue by developing a bevel gear with a curved tooth profile. This type of gear is known as a *spiral bevel gear* (Figure **C**).

C Spiral bevel gear

- **HYPOID GEARS** look very similar to spiral bevel gears, but they are designed so that the axis of the gear does not intersect the axis of the mating gear (Figure **D**). The curved teeth are shaped along a hyperbola (hence the name hypoid gear). Hypoid gears produce much less noise than other gear types and also run more smoothly. They almost always have an axis of rotation 90° to that of the mating gear, and since the shaft axes of both gears do not intersect, it is possible to support both ends of the shafts of both gears. Hypoid gears are commonly used in automotive axles, where a gear reduction and right-angle change of direction are required. The disadvantage to hypoid gears is that their curved tooth profile creates an axial thrust force (similar to helical gears) that must be handled by support bearings. Also, they tend to be relatively expensive.

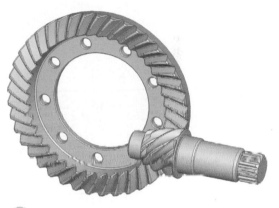

D Hypoid gear set

- A **WORM GEAR** set consists of two gears: a *worm* and a *worm wheel*. In general terms, a worm gear looks like a screw and a worm wheel resembles a spur gear. A worm can have a few teeth, or one long tooth that is wrapped continuously around its base like a screw (Figure **E**). Worm gears are typically used when a large gear reduction is required. The gear ratio of a worm drive is simply the number of teeth of the worm wheel divided

E Worm gear set

BRIAN BUNNELL is a mechanical engineer by education but a maker at heart. He earned his engineering degree from Clemson University in 2000 and has been working in mechanical design ever since. Brian began making early on (creating crazy projects with his dad), and making quickly became his lifelong passion.

SAMER NAJIA holds a degree in mechanical engineering from Duke University but he is a serial maker and building things is his true passion. Samer spends countless hours building progressively larger and more complex projects, disappearing into his garage or loft for hours. Some of these projects are outlandish, but that just means they need more design work.

by the number of teeth of the worm. So, for example, if a worm wheel has 40 teeth, and the worm has only one, then the worm has to rotate 40 times for every one revolution of the worm wheel (40:1), producing a 40:1 gear reduction ratio. Due to the large gear ratios of worm gear sets, they can also transmit a very large amount of torque. For example, when set in an ideal, frictionless state, our 40:1 ratio worm gear provides a torque 40 times greater than the input torque to the worm. Finally, worm gear sets typically cannot be *back-driven*, meaning that the worm wheel cannot drive the worm when a torque is applied to it. The worm gear set can only be moved by rotating the worm.

A winch is a great application of a worm gear set. The high gear ratio and subsequent increase in torque in only two gears makes this an excellent gear choice for a winch. Also, since the worm gear set cannot be back-driven, the winch's load does not fall when no torque is applied to the crank handle (or worm). It is important to note that a worm bears an axial load directly related to the load on the worm wheel. The bearings used to support a worm must be able to handle this axial or thrust loading along with the rotation of the worm.

- **RACK AND PINION SET:** So far, all of the gear types presented transmit only radial or rotational motion. A rack and pinion gear set, however, is used to convert rotational motion into linear motion. The *pinion* is a gear that looks like a regular spur gear. The *rack* is a straight bar with gear teeth cut along one edge of the bar (Figure **F**). In a rack and pinion system, the rack is constrained such that it can only translate back and forth along a linear axis. The pinion gear engages the rack, and when it is rotated, the rack is driven in one linear direction. The rack can be driven in the opposite direction by rotating the pinion in the opposite direction. Car steering uses a rack and pinion gear set to translate the rotational torque from the steering wheel to a linear force applied to the tie rods, causing the wheels to turn.

F Rack and pinion

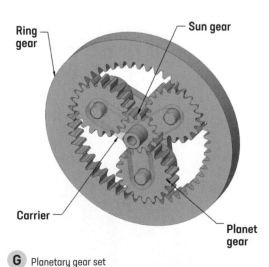

Ring gear

Sun gear

Carrier

Planet gear

G Planetary gear set

Brian Bunnell

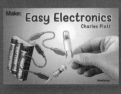

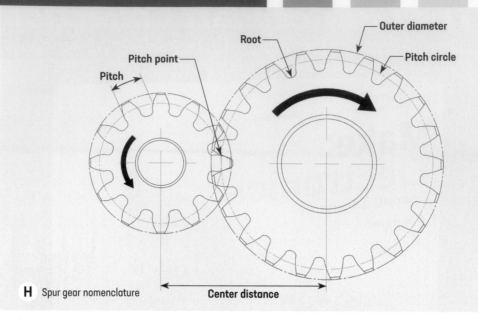

Pitch

Pitch point

Root

Outer diameter

Pitch circle

H Spur gear nomenclature

Center distance

- **PLANETARY GEAR SETS** (also known as *epicyclic gear trains*) contain four main elements. The first is an outer *ring gear* with inward-facing teeth. Engaged with the ring gear are multiple *planet gears*. These planet gears are linked together via a *carrier*. A single, central gear known as the *sun gear* is engaged with all of the planet gears (Figure **G** on the previous page). The ring gear, carrier, and sun gear rotate about the same axis. Each planet gear rotates on an axis constrained by the carrier. As the carrier rotates, the planet gears all rotate about their individual axes while simultaneously "orbiting" about the sun gear. Planetary gear sets are used to transmit a large amount of torque in a relatively small, compact package. They are also able to behave differently, depending upon which part of the gear train is held stationary. They can work as a reduction gear, an increasing gear, or even a reverse gear. This makes planetary gears well suited to applications like automatic transmissions. In fact, this type of gear set is used in many applications ranging from heavy construction equipment to bicycle gear hubs.

General Gear Nomenclature

Before we get into any gear train design specifics and analysis, we need to discuss some general gear terminology. To accomplish this, let's look at our simplest and perhaps most common gear example: the spur gear (Figure **H**).

- **ROOT:** This is the bottom-most part of the gear tooth. The tooth height of two engaging gears is cut such that it is slightly less than the root of the gear. This is so that the tip of each tooth does not hit the root of the gear with which it is engaged.
- **OUTER DIAMETER:** This is a circle describing the outermost extent of a gear. The outer diameter circle lies on the tip of each tooth.
- **PITCH:** The pitch of a gear is the distance between the same point on one tooth and that of the next.
- **PITCH CIRCLE:** The pitch circle of a gear runs roughly through the center of the gear teeth. When properly aligned, the pitch circles of two engaged gears touch at a single tangent point.
- **PITCH POINT:** The point where the two pitch circles of two engaged gears touch. This is the point where the gear teeth make contact.
- **CENTER DISTANCE:** This is a critical part of gear system design that has to be correct for the gears to engage properly. It is the distance between the axis of rotation of one gear to that of a second gear. To find the proper center distance between two gears, add the pitch circle diameter of both gears and then divide by two. ◔

WAYS TO READ Make:

IN PRINT

ON THE GO

ONLINE

Make: is the premier magazine for makers, DIY enthusiasts, hackers, and tinkerers

OVER 120 PAGES OF:
Full project tutorials and skill builders, from electronics and robots to 3D fabrication and woodworking • Maker profiles and interviews • Detailed photos and diagrams to help your builds • Project inspiration • and more!

Subscribe to Make: at makezine.com/go/subscribe
Join Make: Community to access the digital magazine at make.co
Make: Community members and print subscribers have access
online and in the iOS and Android apps.

Adobe Stock-Production Peng

PRUSA SL1 $1,699 prusa3d.com

As you pull the Prusa SL1 out of the box, you know it is different just by the feel. This thing is seriously weighty. That heft comes from the extremely solid construction and large parts that are machined from solid billet. Instead of targeting the super cheap SLA printers available right now as competition, Josef Prusa set his sights on production machines like those from Formlabs. He aimed to create a cheaper machine that was just as reliable and feature-packed. I feel he's done it.

This machine has specs that not only look good on paper, but actually present themselves as useful in practice. The masking screen has an X,Y resolution of 2560×1440 and you can do Z-layers as tight as 0.01mm. It also provides a resin sensor, a

clever tilting bed peel mechanism, and a full-color touch screen. The documentation is thorough and easy to follow. Setup uses touch screen and includes picture examples to ensure that every step is clear.

Operating the unit is as painless as we've come to expect from Prusa products, and the results are fantastic. My prints came out looking shockingly good. And though I didn't have a chance to test it, one feature that makes Prusa printers stand out is customer support. This machine doesn't feel like it is going to break, but I know if by some chance it does, Prusa has great support to back up their products. —*Caleb Kraft*

ROLAND VERSASTUDIO BT-12 DIRECT TO GARMENT PRINTER

$3,495 rolanddga.com

This full-color, high-resolution machine prints directly onto shirts, tote bags, or other fabric goods. With it you can quickly create finished products without a mess, which is an aspect where screen printing falls behind.

If you've used any kind of direct to garment (DTG) printing before, you'll know that the machines are typically cumbersome and large. The BT-12 is relatively compact, smaller than a compact office laser printer. It also comes with an oven that preheats the fabric, then sets the ink for you, all in a small footprint.

This thing is crazy easy to use. The interface is surprisingly simple and I got results that are pretty much photo quality. The prints are supposedly as durable as any product you'd buy in the store. All of this within less than 10 minutes per item.

While the cost of printing individual shirts may be slightly more than using screen printing or similar, if you want something custom and in smaller quantities, it's really difficult to beat this system. The only downside I could find with the BT-12 is that it doesn't have white ink, which means you're limited to printing onto light-colored fabrics. —*Caleb Kraft*

FULL SPECTRUM LASER MUSE 3D AUTOFOCUS

$6,499 fslaser.com/product/muse

The Muse is the latest in Full Spectrum Laser's hobby series of laser cutters and engravers. Its sleek design houses a 40-watt CO_2 laser with a 20"×12" working area.

Retina Engrave, the control software, is easy to use and more than capable of handling pretty much any job. Since the software is hosted on the machine, you can operate the laser with pretty much any computer that has a web browser. As my shop computer died halfway through my tests, I found this to be really beneficial.

The internal RealSense camera worked brilliantly for autofocusing. You can have it focus at the beginning of the job, or, for material with varying surface heights, set multiple points for it to focus on as the laser moves. There's also an option for live autofocusing, which should allow for curved surface engraving, but this needs a little more refinement before I'd feel comfortable recommending it.

Having a camera for placing your designs, and even capturing designs for modification and placement, is my favorite addition to laser cutters in the past few years. I'd love to see CNC machines take notes from this.

—*Caleb Kraft*

FROM OUR BOARDS GUIDE:

Find more board reviews at
makezine.com/comparison/boards

SPARKFUN REDBOARD ARTEMIS

$20 sparkfun.com/products/15444

SparkFun has built four different prototyping boards around the Artemis module, from the smaller Nano board up to the larger ATP (All the Pins) model. In the middle is the RedBoard, SparkFun's Arduino-Uno-like board, now in a version powered by the Artemis module.

There's no one thing about Artemis that jumps out as its killer feature: It's the whole package. New brain on a familiar body? Check. Competitive price, with Bluetooth support as a bonus feature? Check. Stretches a battery? Check. A clear path to mass production for anything you prototype with the board? Check.

A few more nice touches:
» Old USB connectors begone. About time someone made a board with USB-C.
» Built-in microphone
» There's a spot to attach a coin-cell holder, and example programs show how to take advantage of its low power features.
» It's already FCC certified. Manufacture a product built with Artemis and you shouldn't have to pay to do that again.

The RedBoard Artemis feels to me like the Honda Civic of prototyping boards: familiar, well engineered, a bit drab, but in the end oh-so-sensible, and a great value. —*Sam Brown*

RASPBERRY PI 4 MODEL B

$35–$55 raspberrypi.org

For a computer that asks you to plug a keyboard, mouse, and monitor into it, the Raspberry Pi has historically given the feeling you should be able to treat it like a little computer — but stress it too hard and you were never more than one command away from locking it up. So this was the thing I wanted to know: Was the Pi 4 finally the model that would perform like I've always wanted?

I'm happy to say it does. Of everything I've thrown at the Pi 4, I've only seen it crash once, while compiling a particularly hefty bit of C++. And after I stuck a tiny heat sink on it, it ran to completion without a hint of problems.

To my thinking, the Pi 4 is finally the beginner-friendly Pi I've wanted for the maker community all along. It can do everything it promises without needing to go through the more arcane rituals earlier Pis often demanded. No need to re-install a stripped-down version of the OS to free up space and power. No need to tiptoe around, keeping only a few windows open at a time for fear of overloading it. Those days are behind us. The Pi 4 is a hearty little machine in the same tiny size and for the same tiny price we've loved since the beginning. —*Sam Brown*

ARDUINO NANO EVERY

$10–$12 arduino.cc

Like its big brother, the Uno, this board skips all the bells and whistles. Unlike its big brother, the Nano Every fits directly into a breadboard, making it even easier to connect to other parts during prototyping.

That said, what caught my eye about the Nano Every was the header-free version. That means you don't have to use it in a breadboard: It's ready to be more than a prototype. Order your first Nano Every with the headers. Prototype with that one. Then, after you've got your gizmo working, order more Nano Everys sans headers. Whip up your permanent circuit in your CAD program of choice (I like KiCAD), and as you're designing the board, you can copy the exact same connections you made on the breadboard, brain-dead simple. Then send the design off to a manufacturer. When the printed circuit boards arrive in the mail, you can solder the Nano Every straight onto the board, the whole thing all in one go, as if it was just one more part. That option is a godsend for turning out small batches of electronics and getting more copies of your invention into the world faster. —*Sam Brown*

DELUXE FACE SHIELD $25 store.sassafety.com

A while back I saw a tweet from Adam Savage about getting one of the wrap-around face masks that are used by the technicians in the TV show *Westworld*. They looked gorgeous and enticingly futuristic, but were beyond my budget. Thankfully, I found a similarly contoured version that's aesthetically close enough for my liking, reasonably priced (you can find this one at Harbor Freight for just $15), and, most importantly, works great. I've been using it for a couple years now, and it is so much more enjoyable and comfortable than safety goggles. It doesn't fog up on me and gives me a full field of vision, to the point that I often forget I'm wearing it (sometimes only remembering when I try to scratch my itchy nose). My respirator fits underneath it as well. Now I wear it for everything — workshop projects, lawn care, diaper changes, etc. —*Mike Senese*

MAKER'S FIELD GUIDE

$40 amazon.com/Makers-Field-Guide-Definitive-Professional/dp/1732545502/

The *Maker's Field Guide* is positioned as a basic reference for anyone who wants to be able to walk into a workshop and not feel lost. The first few chapters focus on what a workshop is, and what you might find inside, outlining a few different types of workshops like FabLabs and makerspaces. After that, there's a breakdown of some of the most common tools, with a basic anatomical diagram, and some tips for usage and safety. The book finishes off with a few chapters that cover an explanation of common practices such as mold making.

If you're looking for a jump start to get you on your feet before you start building up your real-world experience in the shop, this book may be a perfect fit. —*Caleb Kraft*

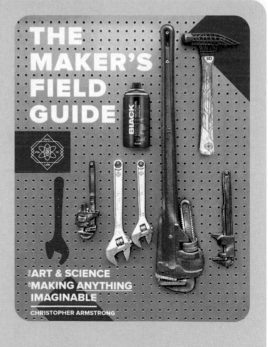

THE MAKER'S FIELD GUIDE

THE ART & SCIENCE OF MAKING ANYTHING IMAGINABLE

CHRISTOPHER ARMSTRONG

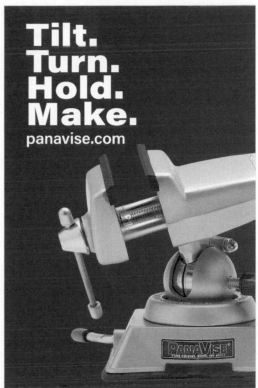

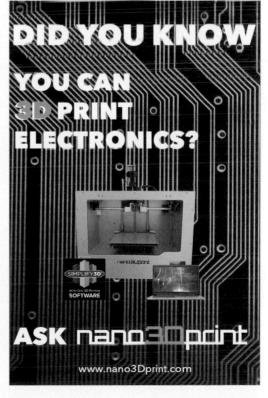

OVER THE TOP

Oklahoma, We Have a Problem

Heather Thomas, Mike Senese

When Heather and Barry Thomas decided to commemorate their fifth wedding anniversary in 2011, they set out to do something a little out of the ordinary.

Rather than exchanging the traditional silverware or wooden gift, the Oklahoma-based couple traveled to a rural road between the nearby villages of Talala and Winganon with a crate of supplies. Their destination: a ditch holding an abandoned cement mixer said to have been too heavy to move after a crash in 1959. Through the decades, locals would decorate the large metal vessel with various motifs, often painted like an American flag. The Thomases had another idea, however, giving the mixer a shiny coat of silver paint, then adding the NASA logo along with hoses, mock rocket nozzles, and other implements to create the semblance of a downed space capsule, lost in the middle of an unlikely prairie.

The result is almost as convincing as it is entertaining, and the once-deserted piece of machinery has now become a must-see roadside attraction for thousands of visitors, complete with its own Facebook page: facebook.com/winganonspacecapsule.

You can find it at E 300 Rd, Talala, OK 74080.

—*Mike Senese*